GW00363602

THE ENCYCLOPEDIA OF
SCIENCE
& TECHNOLOGY

THE ENCYCLOPEDIA OF
SCIENCE
& TECHNOLOGY

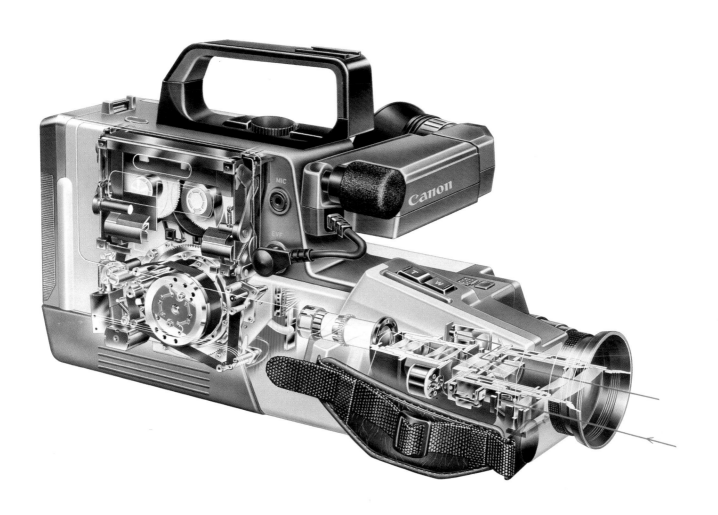

ANDROMEDA

THE ENCYCLOPEDIA OF
SCIENCE & TECHNOLOGY

Consultant Editor: Robin Kerrod
Art Director: John Ridgeway
Designers: David West, Cooper-Wilson
Text Editor: Caroline Sheldrick
Production: Steve Elliott

Media conversion and typesetting:
Peter MacDonald, Una Macnamara and
  Vanessa Hersey

Planned and produced by:
Andromeda Oxford Ltd
11–15 The Vineyard, Abingdon,
Oxfordshire OX14 3PX

Copyright © Andromeda Oxford Ltd 1993

All rights reserved. No part of this
publication may be reproduced or utilized
in any form or by any means, electronic or
mechanical, including photocopying,
recording, or by an information storage
and retrieval system, without permission
in writing from the publisher.

ISBN 1 871869 14 5

Published in Great Britain by
Andromeda Oxford Ltd
This edition specially produced for
Selectabook Ltd

Origination by Hong Kong Reprohouse Co
Ltd, Hong Kong

Printed in Hong Kong by Dai Nippon Ltd

Authors:
Robin Kerrod
Rhys Lewis
Clint Twist

Contents

Introduction

The scientific and technical revolution of the last century has affected every aspect of our lives. We have lasers to play our compact discs. We drive cars made by robots and powered by fuels that are produced by complex chemical processing. Our food is grown using computer calculated amounts of chemicals and our leisure time is often ruled by one of the most successful of all types of mass communication – the television. Most of our machines, from washing machines and televisions to space rockets, are controlled by the silicon chip; although wafer-thin and a few millimetres square, it gives us computer power beyond the wildest dreams of earlier generations.

Each section in *The Encyclopedia of Science and Technology* conveys the excitement and significance of the application of science in our lives. The encyclopedia is divided into clearly defined, self-contained sections, preceded by an introduction to each one. The first of these sections looks at the fuel sources and the ways they are changed into energy that we can use.

Without fuels and other sources of energy, human existence would be cold, dark and primitive. The earliest form of fuel was firewood but today most of our energy comes from fossil fuels: coal, oil and gas. Unfortunately, these forms of fuel are limited and burning them often causes pollution. In recent decades, there has therefore been increased interest in other forms of energy. Until recently, nuclear energy was considered to be the answer, but increasing questions over radioactive leaks and the long-term problems of storage have put its future in doubt.

Consequently, interest is now shifting to energy conservation and renewable sources of energy such as those found in wind, wave power and the Sun's rays. *The Encyclopedia of Science and Technology* looks at all these aspects in detail and uses copious photographs, artworks and diagrams to make the presentation as lively and meaningful as possible.

Power and energy are not the same thing: power is the ability to perform work. The energy from fuels must be transferred and transformed into useful power. Electricity, produced by power stations, is one of the most useful commodities on Earth. Within the last hundred years, the supply of electricity has spread to nearly every part of the globe. As well as providing basic heat and light, electricity powers hundreds of different home appliances and industrial machines. The other vital form of power is motor power, provided by combustion engines. In *The Encyclopedia of Science and Technology*, detailed page by page descriptions show how power stations and how motors work.

For many people, their first interest in technology, and engines in particular, was sparked off by seeing a steam train at work. Great engines powered by steam were the driving force behind the Industrial Revolution of the 19th century. They could be found in all types of industry including mills and factories as well as in traction engines and railway locomotives. In the encyclopedia, the reader can discover how steam engines were invented, and the scientific principles behind the way in which they work.

The steam engine was an external combustion engine; that is, the fuel was burned outside the cylinders. However, it is the internal combustion engine – the petrol engine – that has provided so much of the power for transport this century. The section on power ends with an introduction to jet engines and rocket technology. Although rockets fuelled by gunpowder were invented more than 800 years ago, it was not until the middle of this century that liquid-fuelled rockets were invented. By the early 1960s, large multistage rockets were lifting astronauts clear of the gravitational pull of Earth. The tremendous advances in space technology culminated in Man landing on the Moon in 1969.

Further sections of *The Encyclopedia of Science and Technology* look at the industrial processes involved in working with metals and chemicals. This includes the revolution in the use of plastics; these are now close to metals in being the most important industrial products of our age. Chemicals are also widely used in the growing of crops and in animal

husbandry. The range of foods that are chemically processed has increased enormously since Hippolyte Mège-Mouriès, a French chemist, invented margarine from beef suet, pig's stomach, cow's udder and skimmed milk in 1869. Today, it is possible to buy such things as texturised vegetable protein that looks and tastes just like meat.

The 20th century has continued the revolution in transport which first began with the coming of railways in the 1800s. *The Encyclopedia of Science and Technology* looks at various types of transport from bicycles and cars to vertical take-off and landing (VTOL) aircraft and the F-16 fighter with its speed of more than 2,000 km/h. The reader can discover what the inside of a car looks like and how high-speed trains such as the French Train à Grande Vitesse (TGV) have been developed.

For many people, the biggest impact of modern technology is associated with communications – photography, television, radio, sound recording and, of course, computers. This last section looks at the science and technology behind all these areas from camcorders and digital discs to computer games and the use of robots in industry.

The Encyclopedia of Science and Technology has been designed to deliver vast amounts of information clearly and comprehensively. The rich assortment of images and concepts will enable people of all ages to understand those aspects of science that affect our everyday lives.

Spot facts and fact boxes
Each section in *The Encyclopedia of Science and Technology* is prefaced by a short introduction and a series of memorable "spot facts" which highlight some of the subjects covered in pages that follow. Scattered throughout the book are fact boxes which focus on particular subjects to help reinforce the general text – for example, on how nylon fibres are made or the various different computer languages that exist.

Coal

Spot facts

● In the 30 years 1945-75, we burned as much coal as in all of our previous history.

● The USA has estimated coal reserves of at least three trillion (three million million) tonnes.

● A large bucket-wheel excavator can mine coal at a rate of 8,000 cubic metres per hour.

● The Yallourn Mine in South Australia contains over 13,000 million tonnes of brown coal.

● In a gas explosion in a British mine in 1951, flames travelled 12 km along the underground tunnels.

▶ Shovelling coal deep underground. Although coal mining is now highly mechanized, it still involves a great deal of human labour in very difficult conditions. In some places, coal miners have to work on their hands and knees in tunnels with a roof height of less than 1.5 metres.

Coal is a form of fossilized wood, and coalfields are the remains of great forests that existed hundreds of millions of years ago, during the Carboniferous Period. For the last two hundred years, coal has been our most important industrial fuel, and today it supplies most of our electricity. In countries where large coalfields occur, coal mining is a major industry that uses some of the world's largest machines.

Coal is our most plentiful fuel, and total reserves are enough to last for hundreds of years. Although coal mining damages the environment, and burning coal causes pollution, it is certain to remain a vital source of energy for the foreseeable future.

Fields and reserves

Coal is a black or dark-coloured mineral fuel that consists mainly of the element carbon. As well as carbon, coal contains a variety of hydrocarbons, often traces of sulphur, and moisture. Coal occurs as seams (layers in the rock strata) within the Earth's crust. The depth at which coal seams are found varies enormously. In some coalfields, the seam lies just below the surface. More often, coal seams are buried beneath hundreds of metres of rock.

The properties of coal have been known for at least 2,000 years, but coal was little used until the Industrial Revolution of the 1700s. Before then, firewood and charcoal provided virtually all the world's energy needs. During the 1700s coal quickly replaced charcoal as an industrial fuel. In the 1800s, coal provided the driving energy behind the great Age of Steam. Today, coal is used mainly to produce electricity.

▼ World coal reserves are estimated at around 10 million million tonnes. Most of them are in the Northern Hemisphere, although there are some important deposits in Australia. The USA, USSR and China between them have about 60 per cent of Earth's total resources.

Energy from coal

Total world coal production stands at about 3,800 million tonnes per year. This accounts for just less than one-third (31 per cent) of the planet's total energy production. Coal is still widely used in factory furnaces and domestic stoves, but most of the world's output is burned in power stations to produce electricity. Although it is sometimes considered an old-fashioned fuel, coal is in fact our biggest single source of electrical power, and will long continue to be so.

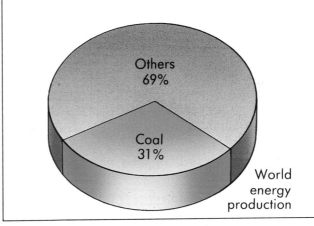

Others 69%

Coal 31%

World energy production

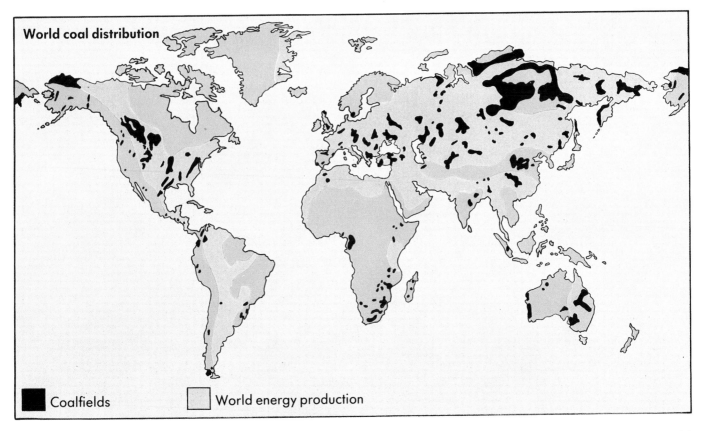

World coal distribution

■ Coalfields   □ World energy production

Coal formation

Coal was formed by the carbonization of trees and plants. When plants die, the carbon in their tissues is normally recycled back into the environment during decomposition. Carbonization occurs when dead plant material is subjected to heat and pressure over millions of years. The different grades of coal were formed by different combinations of time, heat and pressure.

Most of the world's coal was formed during the Carboniferous Period, between 360 and 286 million years ago. At this time, large areas of the Earth's surface were covered with dense swampy forests. Dead plants and trees that fell into the swamps did not decompose completely, but accumulated into thick layers of wet peat. When the swamps were later flooded by the sea, the peat became buried under layers of sediment. Over long periods of time, it decayed further and slowly dried and hardened into brown coal or lignite.

As further layers of sediment built up, increased heat and pressure caused lignite to turn into bituminous coal. In some instances, additional pressures turned bituminous coal into anthracite.

Anthracite and bituminous coal are known as hard coals because of their rock-like appearance. Anthracite is the best quality coal, and contains 86 to 98 per cent carbon. It burns with a bright blue flame and gives off very little smoke. Bituminous coal is the most widely occurring grade of coal, and contains 64 to 86 per cent carbon. Bituminous coal also contains the highest proportion of volatile material, which can be distilled into gas and coal tar. One of the most important discoveries of the Industrial Revolution was the process of baking bituminous coal in an oven to produce coke.

Lignite and brown coals are often referred to as soft coals, and some are soft enough to be crumbled between the fingers. They only contain about 50 to 60 per cent carbon, and are usually compressed into pellets before use.

▼ Fossil remains of the plant *Neuopteris fexuosa* found in coal. The plant lived in the Carboniferous Period (360-286 million years ago) and when the coal-forming material was laid down, it was trapped. Although the plant material decomposed, it left a clear impression in the coal seam.

▲ Coal forms where dead plant material does not properly decay. First it forms peat (1). Once buried and compressed, it becomes lignite (2). Further pressure produces bituminous coal (3) and anthracite (4). The plants were once part of swampy forests in the Carboniferous Period (left).

Nearly coal – peat

The water in swamps and bogs often does not contain enough oxygen or bacteria for normal decomposition to take place. Dead plant material slowly forms a layer of waterlogged peat. Over thousands of years, the layers build up and can reach 30 m in thickness. Although peat is still some way from being coal, it can be used as a low-grade fuel. When it is freshly cut from the ground, peat is a black slimy material containing about 70 per cent water. After it has been dried, it is a crumbly brown solid. When burned, peat gives off large quantities of thick smoke. In many rural parts of Europe, peat is still cut by hand in the traditional manner and is burned in domestic fires. In some countries it is cut by machines and used in small power stations. Most of Europe's peat reserves are now nearly depleted.

Surface mining

When a coal seam occurs just below ground level, it can be worked by surface, or opencast, mining. The overburden, the rocks and soil that cover the seam, is removed, and the exposed coal is mined by mechanical excavators.

In the simplest opencast, or open-pit, method of mining, the coal is dug mechanically out of an excavation. This is used for relatively thick seams with little overburden. For thinner seams, a widely used technique is strip mining, in which the seam is worked by cutting a series of trenches. When all the coal exposed by a particular trench has been removed, another trench is dug alongside and the overburden is used to fill the previous trench.

In Germany's Ruhr valley, the coal seam lies very close to the surface and it is only necessary to scrape aside the topsoil with draglines. In parts of the United States, however, up to 60 m of overburden lie above the coal seam.

In order to be economic, strip mining needs to be carried out on a very large scale. Some of the power shovels used can remove over 150 cubic metres of rock or coal with a single bite.

In general, it is the lower grades of coal that lie nearest the surface, and which are recovered by surface mining. This is especially true in the western United States and eastern Europe where huge deposits of brown coal and lignite are worked. Some high-quality coal is also obtained by strip mining, including almost half the anthracite produced in the United States.

The main problem with surface mining is the damage it creates to the environment, particularly if the coal lies beneath farmland. In many countries, strip mining is now carefully regulated. Some governments require that topsoil is removed and stored separately so that it may be replaced and the area replanted with the minimum of delay.

▲ A giant walking dragline at work in a British coalfield. Draglines of this size require very large and level working surfaces in order to operate efficiently. The ordinary excavator in the foreground of the picture is dwarfed by comparison. It is used to clear obstacles from the dragline's path.

◄ An opencast coal mine in Australia, showing the extensive damage that surface mining can cause to a landscape.

► A power shovel excavator in an American strip mine. Even a very hard coal such as anthracite is soft enough to be worked by mechanical shovels and draglines. Brown coal and lignite are soft enough to be scooped out by bucket-wheel excavators.

Underground mining

▲ Typical winding gear at the pithead, often the only visible feature of a coal mine. This underground mine is in the Rhonda Valley in Wales. Early mines were not very deep and coal was carried to the surface up a series of ladders.

Underground mining is much more widespread than surface mining. Britain, for example, obtains more than 90 per cent of its coal from underground, and mines are often more than 1,000 m deep. Access to the coal seams is obtained by digging a vertical shaft. Mechanical winding gear raises and lowers cages that carry both miners and coal.

Mining is carried out along a series of horizontal tunnels and branching galleries. The exposed portion of the seam that is being worked is called the coalface. Some narrow seams (less than a metre) are still worked by hand. Thicker seams are worked by modern coal-cutting machinery that can cut up to 6,000 tonnes of coal per day.

There are two basic underground coal mining techniques: room-and-pillar and longwall. Room-and-pillar mining, which is still widely used in the United States, involves removing coal from a series of underground rooms. Large pillars of coal are left in place to support the roof of the gallery.

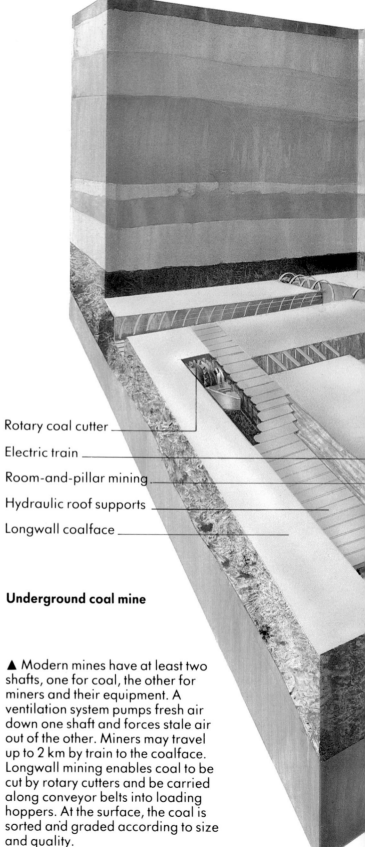

Rotary coal cutter

Electric train

Room-and-pillar mining

Hydraulic roof supports

Longwall coalface

Underground coal mine

▲ Modern mines have at least two shafts, one for coal, the other for miners and their equipment. A ventilation system pumps fresh air down one shaft and forces stale air out of the other. Miners may travel up to 2 km by train to the coalface. Longwall mining enables coal to be cut by rotary cutters and be carried along conveyor belts into loading hoppers. At the surface, the coal is sorted and graded according to size and quality.

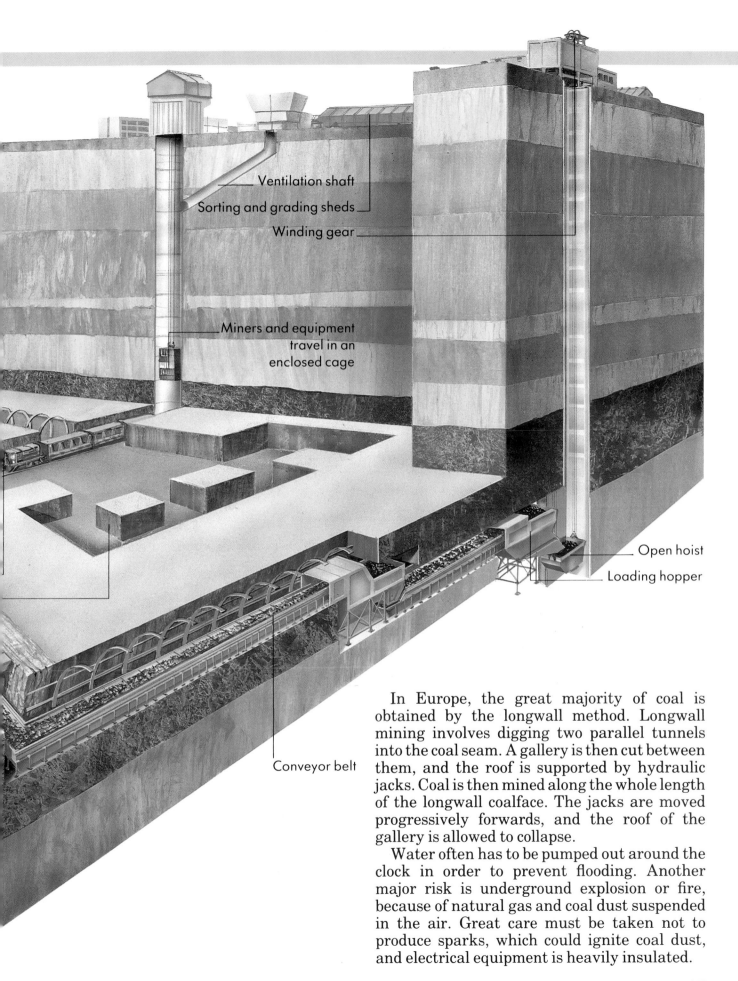

Ventilation shaft

Sorting and grading sheds

Winding gear

Miners and equipment travel in an enclosed cage

Open hoist

Loading hopper

Conveyor belt

In Europe, the great majority of coal is obtained by the longwall method. Longwall mining involves digging two parallel tunnels into the coal seam. A gallery is then cut between them, and the roof is supported by hydraulic jacks. Coal is then mined along the whole length of the longwall coalface. The jacks are moved progressively forwards, and the roof of the gallery is allowed to collapse.

Water often has to be pumped out around the clock in order to prevent flooding. Another major risk is underground explosion or fire, because of natural gas and coal dust suspended in the air. Great care must be taken not to produce sparks, which could ignite coal dust, and electrical equipment is heavily insulated.

Oil and gas

Spot facts

- Virtually all forms of powered transport now use fuels that can be extracted from crude oil.

- The world's largest-diameter oil pipeline carries over one million barrels per day between Texas and New Jersey in the USA.

- More than 40,000 oil-fields have so far been discovered, but fewer than 4,000 of them have any commercial importance.

- The United States has already used up more than half its original oil reserves.

- The world's largest natural gas field is in Siberia, and lies only about 1,200 m below the surface.

▶ Oil production platforms in the North Sea. Small deposits of natural gas that occur in oilfields are normally burned off at the surface in flares. Larger deposits of gas represent a valuable source of energy in their own right. Natural gas is now the world's third most important source of energy.

Crude oil, or petroleum, is the most valuable and the most versatile of Earth's buried treasures. Oil provides us with a number of different fuels, each of which is essential to modern civilization. Petrol, aviation fuel, diesel fuel and heating oil are all refined from crude oil. During the last 100 years, the search for oil has spread to ever more remote and difficult areas: hot deserts, offshore waters and polar wastes.

More recently, natural gas has also emerged as an important fuel, and in many parts of the world it is piped directly into houses for domestic heating and cooking. Oil and gas are frequently found together.

Fields and reserves

By weight, oil and natural gas consist almost entirely of the elements carbon and hydrogen. Chemically, these two elements are combined into thousands of different compounds known as hydrocarbons. Oil and gas occur in natural underground reservoirs that may lie thousands of metres below the surface.

Both oil and gas have been known since ancient times, but they were very little used before the 1860s. The first oil well was drilled in Pennsylvania, USA, in 1859. Within a few decades, oil had also been discovered in other countries. The most recently-developed oilfields are in the polar regions of Alaska and Siberia.

Oil has become the world's most important fuel, which, when refined, is used for domestic and industrial heating and for powering the engines of our machines. Natural gas was once considered a waste product, but is now widely exploited as a heating fuel.

▼ Total world oil reserves are estimated at around 700,000 million barrels. The largest reserves are in the Middle East, which has 26 supergiant fields. Each supergiant field contains at least 5,000 million barrels.

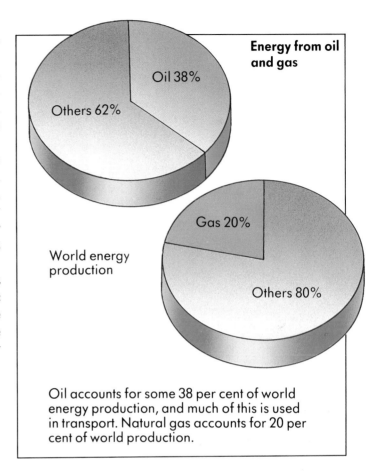

Energy from oil and gas

Oil 38%

Others 62%

World energy production

Gas 20%

Others 80%

Oil accounts for some 38 per cent of world energy production, and much of this is used in transport. Natural gas accounts for 20 per cent of world production.

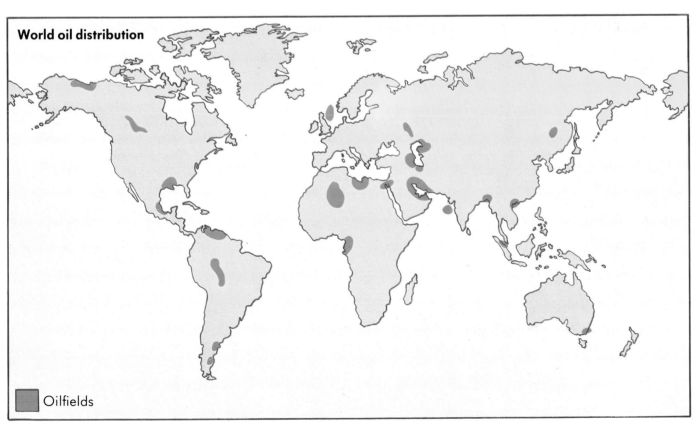

World oil distribution

Oilfields

Formation

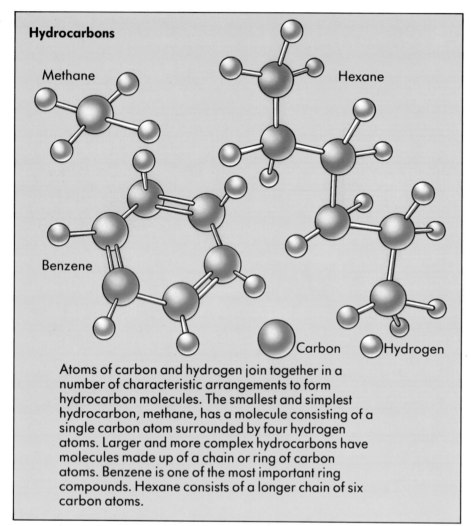

Hydrocarbons

Methane

Hexane

Benzene

Carbon   Hydrogen

Atoms of carbon and hydrogen join together in a number of characteristic arrangements to form hydrocarbon molecules. The smallest and simplest hydrocarbon, methane, has a molecule consisting of a single carbon atom surrounded by four hydrogen atoms. Larger and more complex hydrocarbons have molecules made up of a chain or ring of carbon atoms. Benzene is one of the most important ring compounds. Hexane consists of a longer chain of six carbon atoms.

▲ Pitch Lake in Trinidad, a naturally-occurring deposit that has seeped out on to the surface, forming a lake of liquid bitumen (asphalt). Bitumen is one of the heaviest hydrocarbon compounds.

Oil was formed from the remains of tiny plants and animals that lived in the oceans many millions of years ago. The exact age of oil deposits is hard to determine, because oil moves about easily and is not usually found in the rocks in which it was formed.

At some periods of Earth's geological history, the bodies of algae and plankton accumulated on the seabed and were buried under layers of sediment. The sediment preserved the organic matter from the process of decomposition. Instead, it was transformed by the action of bacteria into a substance known as kerogen. Over long periods of time, further layers of sediment produced increased temperature and pressure, which "cooked" the kerogen and produced many different hydrocarbons. Depending on the exact recipe of the "cooking", crude oil can be thick and dark, or pale and thin.

Oil is chemically stable within the Earth's crust, but a number of factors cause it to migrate physically. As a liquid, oil tends to move through the narrow spaces between particles of rock by capillary action. In rock that is saturated with water, this movement is always in an upward direction because oil is less dense than water. Many oilfields are literally floating on underground water.

Like water, oil passes easily through permeable rocks such as sandstone and limestone, but cannot penetrate impermeable rocks such as slate. Some oil deposits migrate all the way to the surface, but most are eventually trapped by a layer of cap rock. The commonest type of oil trap is an anticline, an arch in the rock strata caused by the folding of the Earth's crust. About 80 per cent of world oil production comes from anticline deposits.

Transporting gas

Before gas can be transported, its volume must be reduced by compression and cooling. Natural gas can then be carried by high-pressure pipelines, or in liquid form on specially-constructed ships. The liquid petroleum gases (LPGs), such as butane and propane, liquefy fairly easily. They can be stored and transported at normal temperatures. Liquefied natural gas (LNG), which is liquid methane, requires constant refrigeration.

▼ (main picture) A section of the 1,300-km long Trans-Alaskan Pipeline snaking across the frozen wastes of Alaska. In the northern part of the state, the pipeline is constructed above ground. This is necessary because the oil passing through is warm and the ground beneath is permanently frozen. Piping the oil through the ground would cause the permafrost to melt, leading to severe environmental damage.

▼ (inset) A team of crawler-tracked pipe-laying machines work in concert to lower a section of the Trans-Siberian Pipeline into a trench. This pipeline carries natural gas 6,000 km from Siberia into Europe.

Refining

Crude oil is of little immediate use, because the individual hydrocarbons are completely mixed together. Useful fuels and other products have to be separated from the crude oil by refining. An oil refinery operates around the clock, processing up to 200,000 barrels of oil per day.

The main refining process is that of distillation, which is based on the fact that different substances vaporize and condense at differing temperatures. Hydrocarbons with small, light molecules, for example methane and butane, are gases at room temperature. The heaviest hydrocarbons, such as asphalt, are almost solid. The liquid fuels: heating oil, diesel, petrol and paraffin, all vaporize at fairly low temperatures.

Simple distillation releases only a small amount of impure fuel. Refineries use a sophisticated process known as fractional distillation to separate crude oil into its component parts, or fractions. The process is carried out inside a hollow steel tower known as a fractionating column. In large refineries these tower more than 45 m above the ground, and contain 30-40 separate condensing plates.

After fractionation, the heavy oils that are left can still be made to produce useful fuel by cracking them. When heated under pressure, their molecules split up into lighter, more valuable hydrocarbons. The cracking process is accelerated if the hot oil is passed through catalysts such as natural and artificial clays.

The most modern refineries use molecular sieves made from certain dehydrated minerals. Depending on the final product required, the broken molecules may then be recombined into other compounds by a second catalytic process. Pumping hydrogen into the catalyst also boosts the efficiency of the cracking process.

Other techniques are also employed. Liquid petroleum gases are sometimes extracted by the absorption process, in which hydrogen is bubbled through the crude oil. The gases are then recovered by washing the hydrogen with steam. Solvents and acids are often used at the end of the refining process to remove impurities.

Refineries also produce white spirits, waxes, greases, and carbon black for printing ink. Many other products are used as feedstock (raw material) for the petrochemical industry.

Fractionating column

◀ Crude oil is heated to about 350°C before it is pumped into the bottom of the column. Oil vapour then rises up the column through a series of steel trays, each of which contains a large number of condensation traps. The trays are maintained at slightly different temperatures, gradually getting cooler towards the top of the column. Different fuels and other products condense in the trays at different heights, and are tapped off. Any gas in the crude oil passes out of the top of the column, and is piped to a gas separation plant.

▶ The lights of an oil refinery glow against the night sky. An oil refinery involves several complex processes. After fractionating, some of the heavier oils are passed to the catalytic cracking plant to be broken into lighter grades. Sulphur may be removed from diesel fuel. The heavier fractions may be further processed, and a vacuum distillation plant produces lubricating oil and paraffin wax.

Nuclear power

Spot facts

• *Five tonnes of uranium fuel will produce the same amount of useful energy as 125,000 tonnes of coal or 500,000 barrels of oil.*

• *Fast, or breeder, reactors actually produce more fuel than they use.*

• *Some waste products produced by nuclear power stations (such as plutonium) will continue to emit dangerous radioactive particles for many thousands of years.*

• *The first nuclear reactor was built by a team under the direction of US physicist Enrico Fermi in Chicago in 1942.*

▶ The control room at the Calder Hall nuclear power station in Britain. Calder Hall was the first commercial nuclear reactor, and it commenced operation in 1956. Today there are about 280 nuclear power stations located in some 25 different countries.

Uranium is the rarest of all naturally-occurring "fuels", and it is also the most powerful. Inside a nuclear reactor, uranium can be made to release energy from the very heart of each atom. Most of this energy is in the form of useful heat. Nuclear power stations use only small amounts of uranium fuel, but produce large quantities of electricity.

Unfortunately, uranium is also our most potentially dangerous fuel. As well as heat, uranium emits potentially harmful radiations. Many people believe that nuclear power production presents a serious threat to the planet Earth. In some countries, the subject of nuclear energy is now highly controversial.

Safe energy?

The first experimental nuclear power plant started operating in the United States in 1951. Initially, there was considerable enthusiasm for this new source of energy. Nuclear power promised to supply the world with large quantities of clean, cheap electricity, especially in countries that lacked reserves of coal and oil. By the early 1960s, more than 100 nuclear power stations had been built. Developing nations were especially eager to acquire this new technology. Today, however, a very large question mark hangs over the nuclear power industry.

The great advantage of nuclear energy is that it uses very little fuel. The great disadvantage of nuclear energy is that it produces high levels of harmful radiation. In 1986, an accident caused an explosion at the Chernobyl nuclear power station in Russia. The explosion created a cloud of radioactive material that contaminated land and therefore food supplies over large areas of Russia and Europe.

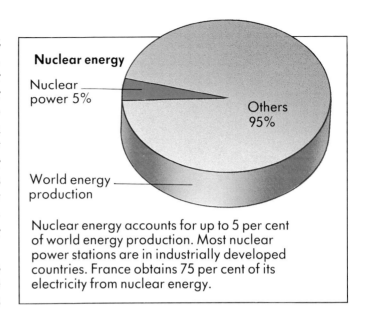

Nuclear energy

Nuclear power 5%

Others 95%

World energy production

Nuclear energy accounts for up to 5 per cent of world energy production. Most nuclear power stations are in industrially developed countries. France obtains 75 per cent of its electricity from nuclear energy.

▼ The nuclear power station at Three Mile Island, Pennsylvania, USA. In 1979, an accident caused the core of the reactor to overheat so much that it began to melt (inset). If the core had melted down completely, huge amounts of radioactivity could have been released, threatening the lives of thousands.

How it works

Nuclear energy is the energy released during the fission, or splitting, of uranium atoms. As well as releasing energy, the fission of a uranium atom also releases neutrons. Some of these strike other uranium atoms, causing them to split, thus releasing more energy and more neutrons. This process is known as a chain reaction. Uranium is the only element which occurs naturally and in which a chain reaction can take place.

Once it has started, a chain reaction tends to accelerate until all the uranium is consumed. Under particular circumstances the chain reaction can be made to happen almost instantaneously. This produces the awesome destructive power of an atomic bomb. Huge amounts of energy are released, but it cannot be put to any constructive purpose.

A nuclear reactor is a device for producing a slow, controlled chain reaction. The energy that is produced is released at useful levels over long periods of time.

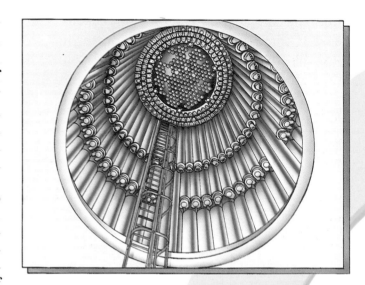

▲ Uranium fuel is loaded into a large number of cylindrical metal containers known as fuel rods. These are then packed closely together to form the core of the reactor. The shape of the core varies with different designs of reactor.

Inside the atom

An atom consists of a nucleus surrounded by shells of orbiting electrons. The nucleus itself is composed of protons and neutrons, held together by an incredibly strong force. When an atom splits, some of this force is converted into very large amounts of energy.

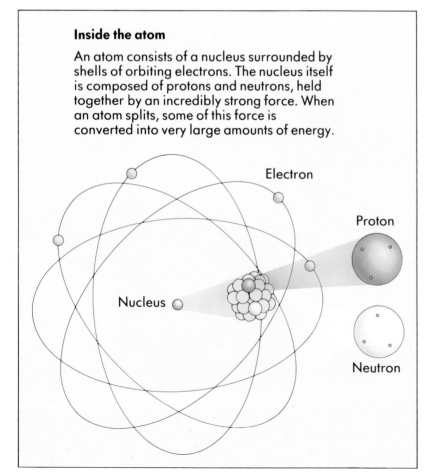

Electron

Proton

Nucleus

Neutron

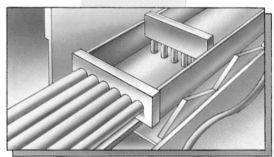

▲ The ore has to be refined in order to produce uranium fuel, which is shaped into thin rods or compressed into pellets.

▼ Mining uranium ore. Even good-quality ore may contain as little as 1-2 per cent uranium. The largest deposits are in North America.

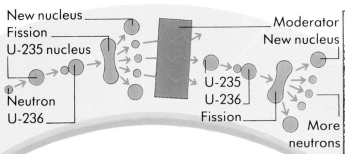

▲ A chain reaction starts with an atom of uranium being struck by a stray neutron. Fission then produces more neutrons, which enable the chain reaction to proceed. The moderator helps control the speed of the chain reaction.

▲ The dangers of radioactivity mean that great care must be taken when transporting nuclear fuel. Special railway trucks are often used. These are designed to withstand a crash at speeds of up to 150 km/h.

▶ All forms of radioactive waste, including the clothing worn by workers, must be carefully sealed and stored. Waste with only a small amount of radioactivity is often stored in steel drums.

▼ Using nuclear fuel requires a cycle of different processes. Uranium must be processed before it can be loaded into a reactor. After use, the fuel must be reprocessed for safe disposal.

◀ Highly radioactive waste must be treated with the utmost care. At this French installation, nuclear waste is being sealed into glass blocks. This reduces the risk of a leakage of radioactivity into the environment.

Reactors

A nuclear reactor has three basic components: a core, a coolant system and a containment. The core produces heat, and the coolant system carries the heat away from the reactor. Most coolant systems operate under high pressure, and the whole reactor is therefore encased in a strong reactor vessel. The containment is an outer covering, usually made of reinforced concrete, that prevents radiation escaping.

The core consists of the uranium fuel rods arranged within a moderator. The moderator serves to slow down neutrons, because slower neutrons bring about fission more readily. The intensity of the chain reaction can be adjusted by a series of control rods made from substances that absorb neutrons. Lowering the control rods causes the chain reaction to slow down. Raising them speeds it up.

The different types of nuclear reactor are designed to make use of different grades of uranium fuel. Natural uranium metal can only be used as a fuel if it is surrounded by an extremely efficient moderator, such as graphite. The uranium is formed into fuel elements, and as many as 30,000 may be stacked into a graphite core measuring up to 15 m high. Such reactors have a coolant system that uses high-pressure carbon dioxide gas, and are usually known as gas-cooled reactors.

The majority of reactors currently in use run on uranium fuel which has been improved by the process of enrichment. Instead of uranium metal, they use a compound known as uranium dioxide. Reactors that run on enriched fuel can use ordinary water, both as a moderator and as a coolant.

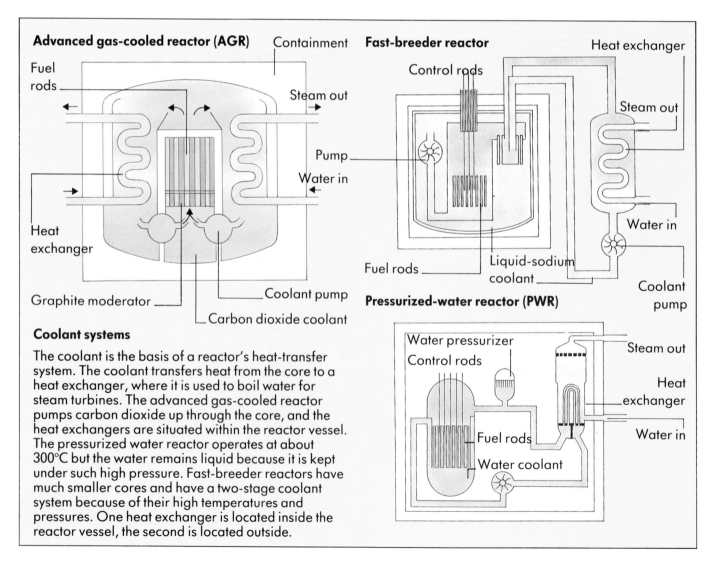

Advanced gas-cooled reactor (AGR) — Containment, Fuel rods, Steam out, Pump, Water in, Heat exchanger, Graphite moderator, Coolant pump, Carbon dioxide coolant

Fast-breeder reactor — Heat exchanger, Control rods, Steam out, Water in, Fuel rods, Liquid-sodium coolant, Coolant pump

Pressurized-water reactor (PWR) — Water pressurizer, Control rods, Steam out, Heat exchanger, Water in, Fuel rods, Water coolant

Coolant systems

The coolant is the basis of a reactor's heat-transfer system. The coolant transfers heat from the core to a heat exchanger, where it is used to boil water for steam turbines. The advanced gas-cooled reactor pumps carbon dioxide up through the core, and the heat exchangers are situated within the reactor vessel. The pressurized water reactor operates at about 300°C but the water remains liquid because it is kept under such high pressure. Fast-breeder reactors have much smaller cores and have a two-stage coolant system because of their high temperatures and pressures. One heat exchanger is located inside the reactor vessel, the second is located outside.

Most water-cooled reactors use water at over 100 times atmospheric pressure, and are known as pressurized-water reactors (PWR). Other designs make use of boiling water as a coolant.

If uranium fuel is very highly enriched, it can be used in reactors that do not need a moderator. This type of reactor is known as a fast reactor. Fast reactors can also make use of plutonium, an element extracted from depleted fuel. One major advantage of fast reactors is that they can be used to "breed" more plutonium from depleted uranium fuel. For this reason they are often called breeder reactors.

Fast reactors operate at higher temperatures than other reactor types, and produce energy more efficiently. There are several designs, and all of them use liquid sodium as a coolant.

▼ The chart shows the electricity generating capacity of nuclear power stations ordered by European countries during the period 1956-85. Nuclear power became most popular when oil prices rose during the 1970s. Since the accident at Three Mile Island in 1979, nuclear energy has undergone a worldwide decline in popularity. This is clearly indicated by the number of new reactors ordered in Europe after 1980.

Hands off

Even small doses of radiation can be harmful to human health. Workers at nuclear power stations take every precaution to prevent their exposure to radiation. Inside the reactor building, workers wear heavy protective suits lined with radiation shielding. Delicate operations, such as removing depleted fuel rods, are usually carried out by remote control from behind heavily shielded walls.

Commercial orders for nuclear reactors

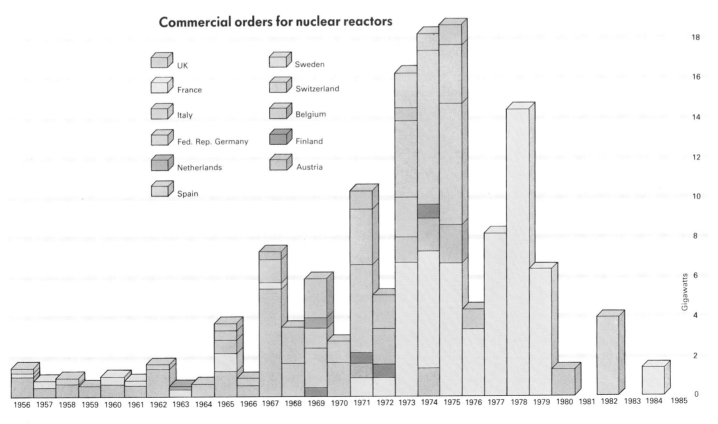

UK
France
Italy
Fed. Rep. Germany
Netherlands
Spain
Sweden
Switzerland
Belgium
Finland
Austria

Solar energy

Spot facts

• The total energy contained in sunlight reaching the Earth is equivalent to the output of about 200 million nuclear power stations.

• At least 60 million solar-powered calculators have now been sold worldwide.

• The first house to rely completely on solar heating panels was built in 1939 at an American university.

• Driven solely by solar power, the US-designed vehicle Sunraycer achieved a speed of over 78 km/h in 1988. In the previous year it had won a 3,140-km race for solar cars across Australia.

▶ Located about 150 million km away, the Sun is a huge nuclear furnace that radiates vast quantities of energy into space in all directions. Only a very small proportion (about a thousand millionth) of that energy reaches Earth.

The Sun represents an inexhaustible source of free energy. Most buildings already make some use of passive solar heating, and in many countries the Sun's energy is actively collected to provide hot water for household purposes.

The main drawback with solar energy is that it produces only low temperatures under natural conditions. In order to produce useful quantities of electricity from solar energy, the heat energy in sunlight must be collected over a large area and concentrated at a single point. In countries with suitable climates, experimental solar-energy power stations are in operation.

Sun power

Solar energy dwarfs all our other energy sources. In less than one hour, the Earth receives energy from the Sun that is equivalent to the world's total energy output from other sources during an entire year.

Most of the Sun's energy is reflected back into space or is absorbed by the atmosphere. However, sunlight still reaches the Earth's surface in usable quantities. On a summer's day in Britain, for example, the energy falling on one square metre of sunlit ground is equivalent to ten 100-watt light bulbs.

Sunlight provides a constant source of energy for the Earth as a whole, but it is not evenly distributed over the planet's surface. Solar energy can only be exploited where and when the sun is shining. Usually this means countries with hot climates and clear skies. Even in cold northerly countries, however, solar energy can still be very useful.

The energy in sunlight can most easily be exploited in the form of direct heat. Rooms can be heated simply by letting the Sun shine in freely. Higher temperatures, needed to provide domestic hot water, require the active technology of solar collection panels.

Sunlight can also be converted directly into electricity using solar cells. At present, these are mainly used on Earth in calculators and watches that consume only small amounts of electricity. But larger solar-powered devices also work: a solar-powered aircraft has flown between Britain and France; and solar-powered cars have been built in several countries.

▼ Some 30 per cent of the Sun's energy that reaches Earth is reflected back into space by the atmosphere. Virtually all of the remaining 70 per cent is absorbed by the atmosphere, where it powers the water cycle. Direct heating of the Earth's surface, which causes winds and currents, accounts for less than 1 per cent.

Solar energy

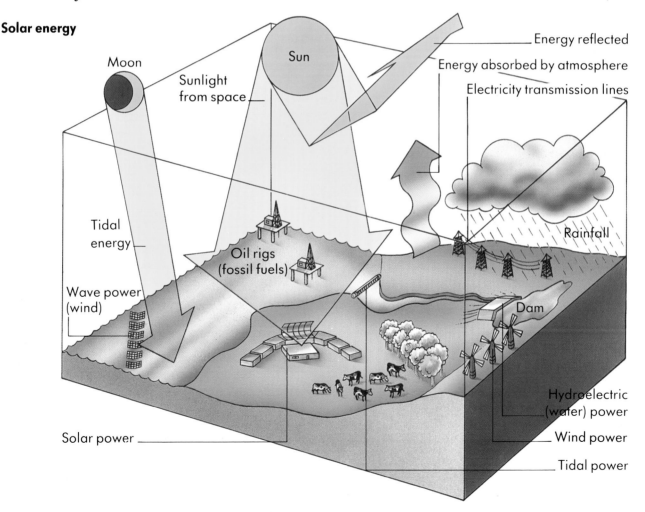

At home

All houses and other buildings already make use of solar heating from sunlight shining on walls and rooftops, and through windows. Even in countries which have cool, cloudy climates, houses obtain as much as 20 per cent of their space heating (room heating) requirements from the Sun. By incorporating passive solar technology, such as sloping windows and trombe walls, this can be boosted up to 80 per cent. This sort of heating is known as solar gain. Houses that are designed to make maximum use of solar gain are usually well insulated in order to keep heat loss to a minimum.

Producing domestic hot water from the Sun requires an active solar collector. The commonest type is the flat-plate collector, which consists of a coil of water-filled hosepipe inside a sealed glass-topped box. Black pipe is used for the coil because dark-coloured surfaces absorb heat better than lighter ones. The sealed box acts in the same way as double glazing, and reduces heat loss. The water in the pipe is connected to a closed system, and heat is transferred through another coil to a hot-water cylinder. This is more efficient than using the Sun-warmed water directly from the pipe.

Millions of these roof-top solar collectors are now in use worldwide, particularly in the countries around the Mediterranean Sea.

▼ This British house makes good use of solar gain. Large sloping windows are the simplest form of solar technology. A sloping surface can receive up to 10 per cent more solar energy than a vertical one. This principle is used in many solar-energy devices.

▲ This American house design uses light-sensitive switches to retain the heat obtained from solar gain. When the Sun is shining, the windows open and air can circulate through the house. When the Sun goes in, the switches close the windows to trap the heat.

In countries which receive less sunshine, vacuum-tube collectors can be used to supply houses with hot water. A vacuum-tube collector consists of a black metal collection plate fitted inside a sealed glass tube. A vacuum inside the tube insulates the plate against heat loss. Heat is transferred by a coolant liquid circulating within the collection plates. A typical installation may contain 20 or 30 tubes connected together.

Vacuum-tube collectors are more expensive than flat-plate collectors, but can deliver about twice as much useful energy over a year. By surrounding the tubes with curved mirrors, or by focusing the Sun's rays through lenses, higher temperatures and greater efficiency can be achieved.

Active and passive

The flat-plate collector is a form of active solar technology. The Sun's energy is used to heat water, and that heat is then transferred to a separate supply of water. A trombe wall is a form of passive technology that makes greater use of solar gain. Sunlight shining through the glass outer wall is absorbed by the dark-coloured inner wall. Convection currents circulate warm air into the room, and draw cold air out. The inner wall also radiates heat into the room.

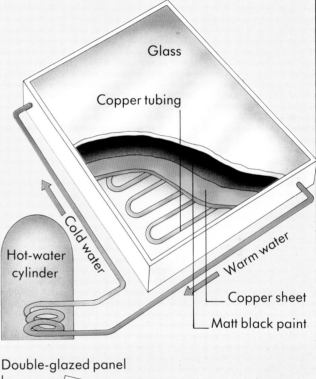

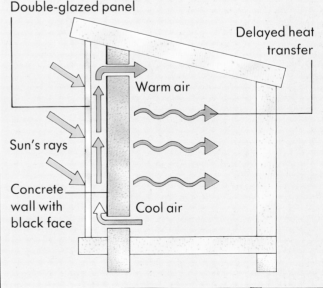

Solar electricity

Producing useful quantities of electricity from solar energy requires a very large-scale installation. The most widely-adopted design is that of the solar field. This is already being used in several countries including Australia, Japan, Spain, Italy and the USA.

A solar field consists of many rows of individual solar collectors. These are connected to a central heat exchanger that produces steam to drive an electric generator. The collectors are normally surrounded by curved reflectors, and are made even more efficient by the process of Sun-tracking. Each collector is mounted so that it can be swivelled and tilted to be always facing directly at the Sun. Throughout the day, the position of the collectors are constantly adjusted by small computer-controlled motors.

The main disadvantage with the solar field is that heat energy is lost during the transfer from the collectors to the central heat exchanger. One solution to this problem is to concentrate the Sun's heat into a central collection point by using a circular field containing thousands of separate mirrors. The first central collection systems were experimental solar furnaces that could reach temperatures above 3,000°C. During the last ten years, however, the first central collection power towers have begun operating. The Sun's rays are focused at the top of the power tower, and heat is collected by a series of black-coloured pipes containing liquid sodium. Heat exchangers at the base of the tower are connected to boilers that produce steam to drive generators.

Photo-voltaic cells can also be used to produce large quantities of electricity, but at present the process is too expensive to be practical. They are more efficient in space than on Earth.

▶ The world's largest power tower at Barstow, California, USA. The tower itself stands about 90 m tall, and the field of mirrors covers some 90,000 square metres.

Solar cells

A solar cell, often called a photovoltaic (PV) cell, converts the energy in sunlight directly into electricity. An individual cell consists of two thin slices of silicon crystal sandwiched between two layers of metal. The top layer of metal is in the form of a grid so that sunlight can reach the upper side of the silicon. The two slices of silicon contain slightly different amounts of impurities, causing them to have different electrical states. Sunlight falling on the upper slice causes electrons to flow into it from the lower slice. This creates an electrical current that flows through the metal contacts. The photo shows panels of solar cells mounted on a research satellite. In space, the cells will receive the full strength of the Sun's rays, and will be able to operate at maximum efficiency. On the Earth's surface, however, even strong sunlight has had most of the energy filtered from it by the atmosphere.

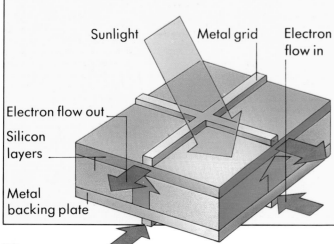

Sunlight | Metal grid | Electron flow in

Electron flow out

Silicon layers

Metal backing plate

Nature's power

Nearly all of nature's power that we harness comes in the form of movement. It is largely a case of converting one sort of movement into another. The movement of running water and the rushing wind is converted into rotation by machines based on the wheel.

A waterwheel is placed edge-on into a flow of water. The wheel is turned by the force of the flow against blades set across the wheel's rim.

In a traditional windmill, the wheel takes the form of a number of angled blades, or sails, which are placed face on to the wind. A windmill turns because the wind is deflected by the angled blades as it flows through the wheel.

Waterwheels and windmills have been in widespread use for at least 2,000 years. Since Roman times they have provided useful energy for milling grains for flour, or for pumping water for irrigation and drainage. At the beginning of the Industrial Revolution, water power provided most of the energy that drove the spinning wheels and other machinery in the first factories.

During the last 100 years, the energy conversion process has been taken one step further. Water and wind power are now used to drive turbines. These in turn are used to generate energy in the form of electricity.

▶ Traditional windmills provided a steady source of low-speed rotation that was very useful for certain tasks such as grinding corn or pumping water.

▼ An ancient waterwheel in Syria, possibly dating from Roman times. Waterwheels could be almost any size, but until the 1700s the materials used were very weak.

Water power

Water is the most useful source of natural power because it is the easiest to control. Running water can be channelled along sluices and through pipes. More importantly, a river can be blocked by a dam, creating a reservoir that can store huge quantities of water. Water from the reservoir can then be released as and when it is required.

Water power is harnessed to generate electricity in hydroelectric power (HEP) stations. These are usually situated at the base of a large dam. The best locations for HEP projects are the narrow, steep-sided river valleys found in mountainous areas. A dam across such a valley can create a reservoir more than 100 km long. Large-scale HEP projects may involve more than a simple dam and reservoir. In the Snowy Mountains of Australia, the waters of the Snowy River have been diverted by a series of underground tunnels to some 16 HEP stations.

Water power can also be used to store surplus energy from other power stations. This is carried out in what are known as pumped-storage HEP schemes. These use two separate reservoirs at different levels.

During normal operation, water from the upper reservoir is used to drive turbines to produce electricity. After passing through the turbines, the water is stored in the lower reservoir. Whenever there is a surplus of electricity, it is used to pump water from the lower reservoir back up into the higher one. Demand for electricity is at its highest during the day. This means that, in most stations, pumping is often done at night.

▼ Construction workers inside the tunnels at a HEP station. The worker on the right is in the main water-supply tunnel. The worker on the left is standing at the mouth of an intake pipe leading to a turbine.

Hydroelectricity

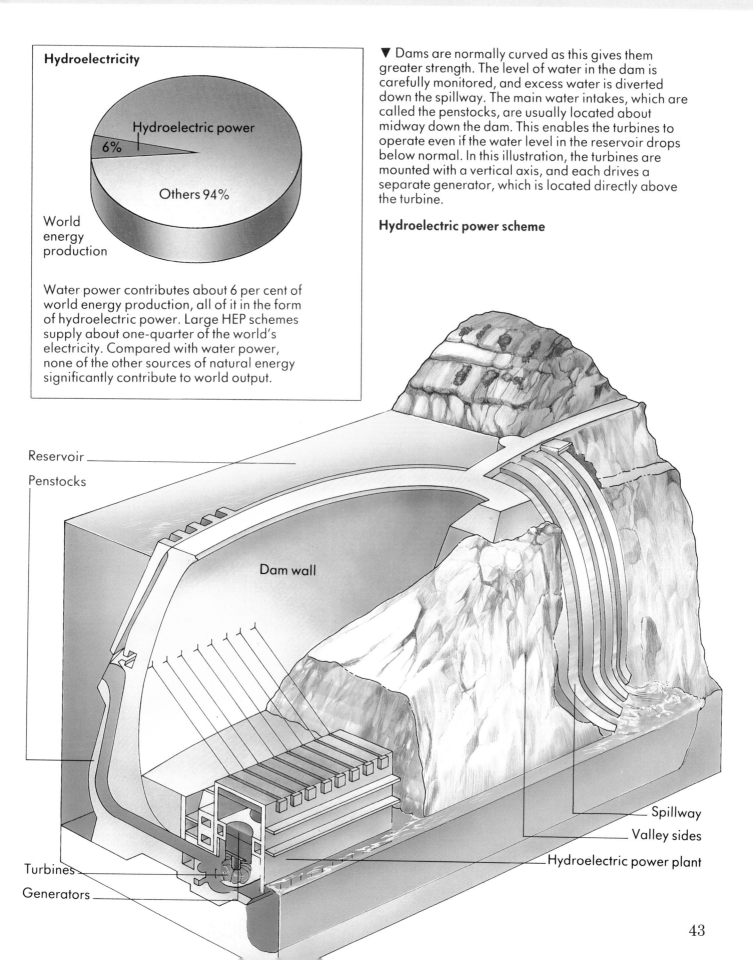

Hydroelectric power

6%

Others 94%

World energy production

Water power contributes about 6 per cent of world energy production, all of it in the form of hydroelectric power. Large HEP schemes supply about one-quarter of the world's electricity. Compared with water power, none of the other sources of natural energy significantly contribute to world output.

▼ Dams are normally curved as this gives them greater strength. The level of water in the dam is carefully monitored, and excess water is diverted down the spillway. The main water intakes, which are called the penstocks, are usually located about midway down the dam. This enables the turbines to operate even if the water level in the reservoir drops below normal. In this illustration, the turbines are mounted with a vertical axis, and each drives a separate generator, which is located directly above the turbine.

Hydroelectric power scheme

Reservoir

Penstocks

Dam wall

Turbines

Generators

Spillway

Valley sides

Hydroelectric power plant

Wind power

Unlike water, the wind cannot be controlled or stored. Wind power must be exploited where and when it occurs naturally. Until very recently, the wind was mainly used to drive small pumps for agricultural purposes.

The amount of power produced by the wind increases as the cube of its velocity. This means that a doubling of the wind speed produces eight times as much power. In general, wind speeds increase with altitude. At 10 m above the Earth's surface, the wind speed is about 20 per cent greater than at ground level. At 60 m up, velocities may be 50 per cent greater.

Traditional windmills were designed to operate at fairly low wind speeds. The materials they were made from (wood and cloth) were not strong enough to withstand high winds.

Modern windmills, which are usually called wind turbines, are designed to operate at much higher velocities. As a result, they produce far more power, and can be used to generate electricity. There are two main types of wind turbine. The horizontal-axis turbine has the same basic layout as a traditional windmill. Instead of sails, it has a rotor shaped like an aeroplane propeller.

▼ Small wind turbines are used throughout the world to pump water and generate small amounts of electricity, particularly on farms. The commonest design uses a rotor consisting of a large number of metal vanes. The rotor turns on a horizontal axis. The whole of the turbine assembly is on a swivel mounting so that it can be turned into the wind by the attached rudder.

▶ A wind farm in California, USA, consisting of many rows of small Darreius wind turbines. Darreius turbines are easily recognizable by their distinctive shape, and can operate in wind coming from any direction. This particular wind farm is situated in a high mountain pass, where the winds are unusually strong and steady. Careful siting is the key to wind farming.

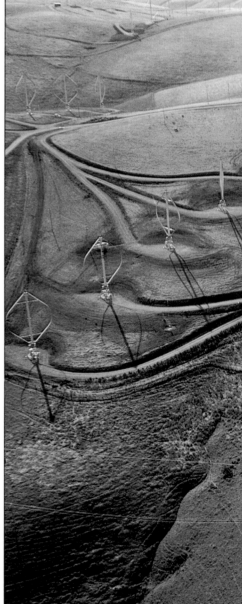

A vertical-axis wind turbine rotates on a shaft that is vertical to the ground. It normally has only two blades, mounted vertically at each end of a horizontal rotor.

The Darreius wind turbine is an advanced vertical-axis design that uses two curved blades. There is no separate rotor, and the blades are attached at each end of the shaft.

Wind turbines of both basic types are now in operation in many countries throughout the world. The largest ones are more than 100 m tall. At full speed, the tips of the blades travel at up to 400 km/h.

Wind turbines are positioned wherever the winds are strongest, and are often located on hills and clifftops. As with solar power, there are two main approaches to the large-scale use of wind power. A wind farm is a large area of land containing many small wind turbines, up to 30 m tall. Each of these contains a separate electrical generator. The other approach is to build just one or two very large turbines at each location. Taking maximum advantage of wind power may mean building wind farms offshore, where wind speeds are generally higher than over land.

Geothermal power

The Earth's crust is the thin, solid outer layer of our planet. Below about 30 km, heat from natural radioactivity is sufficient to keep rock in a molten state. The temperature rises steadily with depth, generally about 30°C for every 1,000 m of depth. In areas of volcanic activity, this can increase to 80°C per 1,000 m, and higher temperatures occur much closer to the surface.

In a few rare instances, in the USA, Japan and Italy, this heat boils underground water, which rises to the surface as dry steam. This steam can be trapped and used to drive turbines. In California, the Geysers power station has been built on top of a vast underground reservoir of dry steam. When operating at full capacity, the Geysers power station supplies nearly all the electricity required by the city of San Francisco.

In most cases, superheated water remains trapped underground. When brought to the surface by wells, the water boils and the steam is used by turbines. Several countries, as far apart as Mexico and the Philippines, already produce electricity in this way.

Low-temperature geothermal heat has been used for thousands of years in public baths and health spas. In some parts of the world, hot water from volcanic springs and geysers is now a major source of domestic heating. In Iceland, more than two-thirds of the population now heat their homes with natural hot water. Other countries that make use of geothermal heating include the USA, Russia, China, Japan, France, New Zealand and Hungary.

Even when there is no naturally-occurring underground water of suitable temperature, geothermal energy can still be harnessed. Hot, dry rocks may be used soon in many countries as a huge underground boiler.

▼ (right) The Geysers geothermal power station in California, USA, supplies electricity to a city of half a million people. (left) This design is for a power station producing electricity from hot dry rocks. Two wells are drilled some distance apart, one deeper than the other. The surrounding rock is then fractured with explosives to produce a large number of heat transfer surfaces. Cold water is pumped down the deeper well into the fissured rock, where it boils. Steam is tapped off by means of the other well.

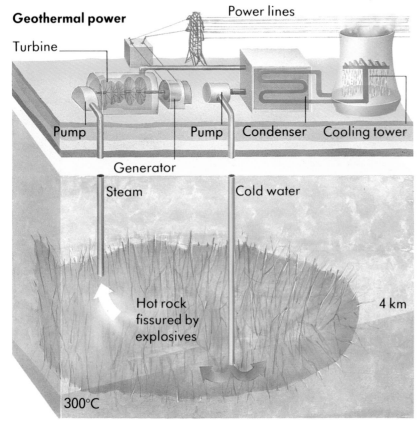

Geothermal power

Turbine · Power lines · Pump · Pump · Condenser · Cooling tower · Generator · Steam · Cold water · Hot rock fissured by explosives · 4 km · 300°C

Tidal and wave power

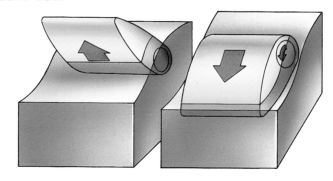

◄ A tidal-power barrage across the estuary of the River Rance in France. At 750 m long, the barrage forms the world's largest tidal-power station and contains 24 turbines, which can operate when the tide is flowing in either direction.

▼ Wave energy converters will need to be very large. (1) The Salter duck uses a string of floating ducks hinged on a common shaft; a full-sized duck would measure 25 m across. (2) A development of the air-bag principle. The motion of the wave pumps air from one side of the converter to the other. The estimated length is 250 m. (3) Each wave-contouring raft would cover at least 5,000 square metres.

1 Salter duck

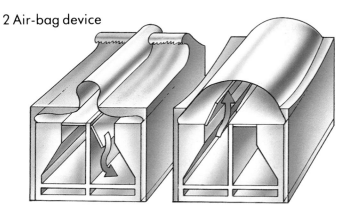

2 Air-bag device

3 Wave-contouring raft

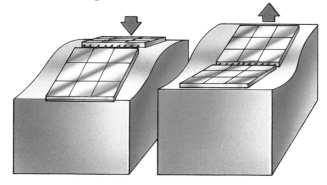

When the motion of the tides is channelled by a natural feature such as a river estuary, it produces a very strong flow of water. The power of the tides can then be harnessed by building a barrage fitted with turbines across the tidal flow. Tidal-power barrages have been built in France, Russia and China.

The ceaseless up-and-down motion of ocean waves also represents a potentially valuable source of energy. A number of wave-power devices have been invented and tested as models, but none of them yet operate.

There are two basic approaches to wave power. The simplest designs place a line of wave-energy converters across the path of incoming waves. The energy converters float on the surface and consist of two sections hinged together. The motion of the waves operates the hinge, which powers pumps that drive turbines.

Other designs place the energy converter edge-on to the waves. The converters contain a number of air bags. As the wave travels along the converter, it squeezes the bags and forces air through turbines.

Power stations

Spot facts

• *The largest and most powerful steam turbines can each produce enough electricity to supply nearly 800,000 houses.*

• *The condensers for the steam turbines of a 2,000-megawatt power station require more than 200 million litres of cooling water every hour.*

• *Fuel-burning power stations have an average efficiency of about 35 per cent: nearly two-thirds of the heat energy they produce is wasted.*

• *The hydroelectric power scheme on the border between Brazil and Paraguay produces enough electricity to supply a city of as many as 35 million people.*

▶ Tall and unsightly cooling towers are a characteristic feature of many power stations. They recycle cooling water for the condensers on the steam turbines. Hot water from the condensers is piped to the top of the towers and sprayed downwards. By the time the water reaches the ground, it has cooled enough for it to be reused.

Electricity is produced by the constant high-speed rotation of turbine-driven generators. The story of electricity is also a story of turbines. The widespread use of electricity only became possible with the invention of the steam turbine in the late 1800s.

Today, power stations fitted with steam turbines generate about three-quarters of the world's electricity. The remainder is produced by water turbines in hydroelectric power stations. During the last ten years, there has been considerable interest in wind turbines as a source of electricity. This kind of power generation is less harmful to the environment. Many experimental designs have been built, but so far they make no significant contribution to world energy production.

The generation game

The useful properties of electricity were first demonstrated by the British physicist Michael Faraday. In 1821, Faraday built the first electric motor. Ten years later, he developed the principle behind the electrical generator, which is basically an electric motor in reverse. By the 1860s, small electrical generators powered by steam engines were being used in many parts of Europe and the United States.

The first generating stations were built to supply large individual buildings, such as hospitals and indoor markets. During the 1880s, demand for this new source of energy grew rapidly, and the first central generating stations were built. These were intended to supply electricity to the general public across a whole city district. In 1882, the American inventor Thomas A. Edison opened the famous Pearl River power station in New York.

In order to produce electricity in commercial quantities, generators need to rotate at speeds of at least 1,000 rpm (revolutions per minute). Steam engines could only just achieve that speed, and it meant operating to the limits of their capacity.

During the 1880s and 1890s, the British engineer Charles Parsons perfected the inward-flow steam turbine that could produce speeds of up to 18,000 rpm. Parson's invention was rapidly adopted by the new electrical industry, and the first large steam turbines were installed in a German power station in 1901. Since then, steam turbines have become our single most important source of electrical power.

Most power stations produce electricity in a two-stage process. Fuel of whatever kind, whether it be coal, oil, gas or uranium, is first used to produce steam. The steam is then used to drive a turbine, which in its turn produces useful quantities of electricity.

▼ The generator room at the Paris Opera House, installed by the American Thomas Edison in 1887. Steam engines worked flat out to provide high speeds of rotation. The rotary power was transferred to the generators by a series of belts. Today's power stations have turbines and generators mounted in line, and sharing a common rotating shaft.

▼ The first electrical generator built by Michael Faraday. Electrical current was produced by a copper disc spinning between the poles of an electromagnet. Modern generators use coils of copper wire instead. The first generators produced direct current. The switch to alternating current was made during the 1890s.

Steam turbines

A steam turbine consists of a central shaft, or rotor, mounted horizontally within a cylinder. The outer surface of the rotor is fitted with a large number of angled blades, which radiate like the spokes on a wheel. High-pressure steam is passed into the cylinder through a series of nozzles, mounted around its inner surface.

As the steam enters the cylinder, it expands. The heat energy of the steam is the kinetic energy (the energy of motion) of the vapour molecules. This energy is transmitted to the shaft through the angled blades, thus causing it to rotate. The speed of rotation depends on the temperature and pressure of the steam.

The introduction of new materials has enabled the construction of steam turbines that operate at extremely high temperatures and pressures. The design of the angled blades is also an important factor.

The simplest form of steam turbine is the impulse turbine, in which the blades are shaped like tiny cups. The steam releases all of its energy when it hits the turbine blades. The shaft rotates, but the steam itself comes to a dead stop.

A reaction turbine, of which Parson's original turbine was an example, allows the steam to flow through a series of fixed and moving blades. These are designed to allow the steam to continue expanding as it passes through them. This means that the steam energy is used more efficiently, and the flow of steam is twice as fast as that of an impulse turbine. Modern steam turbines usually incorporate both types of blades in their design.

Still greater efficiency can be obtained by allowing the energy transfer to take place in a number of separate stages rather than all at once. This process, which is known as staging, is now used on all steam turbines.

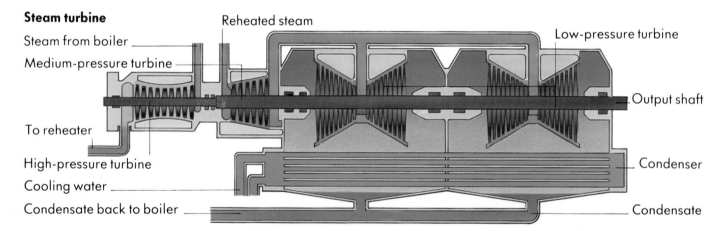

Steam turbine

Reheated steam

Steam from boiler

Medium-pressure turbine

To reheater

High-pressure turbine

Cooling water

Condensate back to boiler

Low-pressure turbine

Output shaft

Condenser

Condensate

▲ In a modern steam turbine, superheated high-pressure steam expands through progressively larger turbine blades. It is piped first into the high-pressure turbine, and, after reheating, into first the medium-pressure and then the low-pressure turbines. The steam condenses back into water in the condenser and is returned to the boiler.

▶ Steam turbines being assembled in a factory. In order to withstand constant exposure to high-temperature steam, the sets of blades have to made from extremely tough and resistant materials.

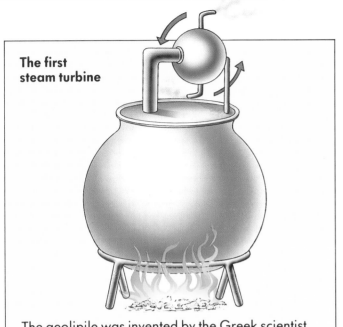

The first steam turbine

The aeolipile was invented by the Greek scientist, Hero of Alexandria, about 1,900 years ago. Jets of steam caused the ball to rotate.

The commonest method of staging is to allow the steam to pass through a series of angled blades of progressively larger diameters. The steam drives the smallest set of blades first.

Large turbines take staging a step further by incorporating two or three separate sets of turbine blades mounted on the same shaft. Each stage is contained within a separate cylinder. The first stage usually has impulse blades and uses very high pressure steam.

After the steam has passed through the first stage, it is collected and reheated until it has sufficient energy to drive medium-pressure turbine blades. The steam is then piped into the low-pressure turbines. Turbines like these are known as compound turbines.

▼ A large compound steam turbine of the type found in most power stations. The many sets of steam-driven blades are all attached to one central shaft.

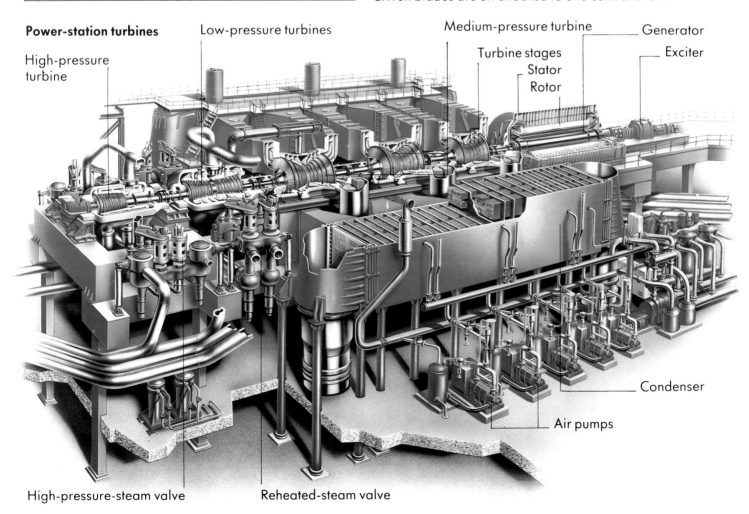

Power-station turbines

High-pressure turbine

Low-pressure turbines

Medium-pressure turbine

Turbine stages
Stator
Rotor

Generator

Exciter

Condenser

Air pumps

High-pressure-steam valve

Reheated-steam valve

Nuclear power

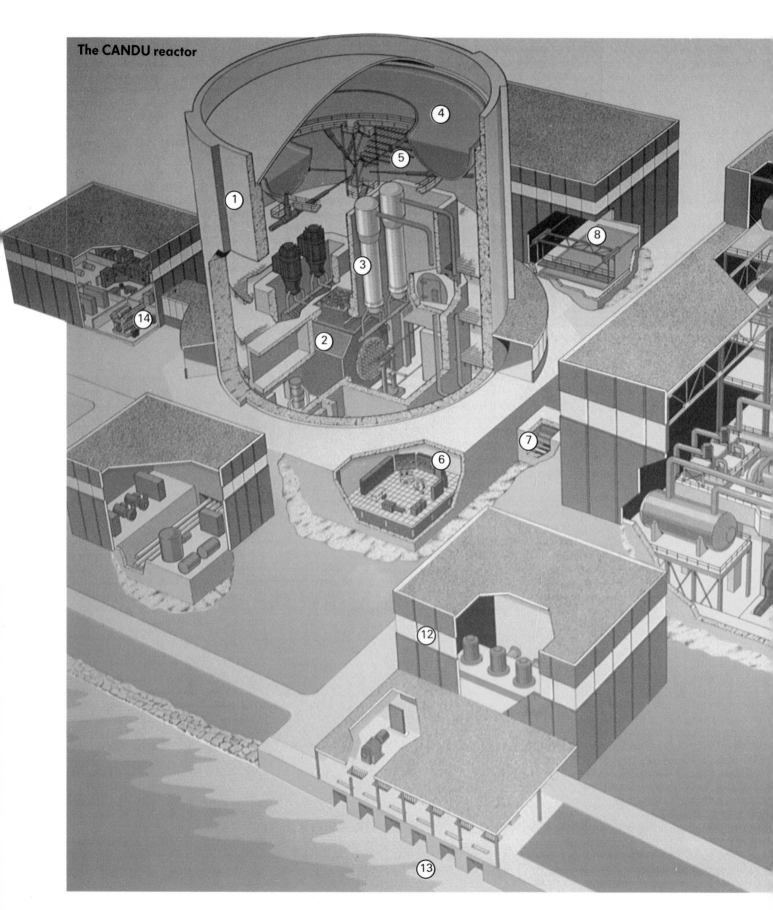

The CANDU reactor

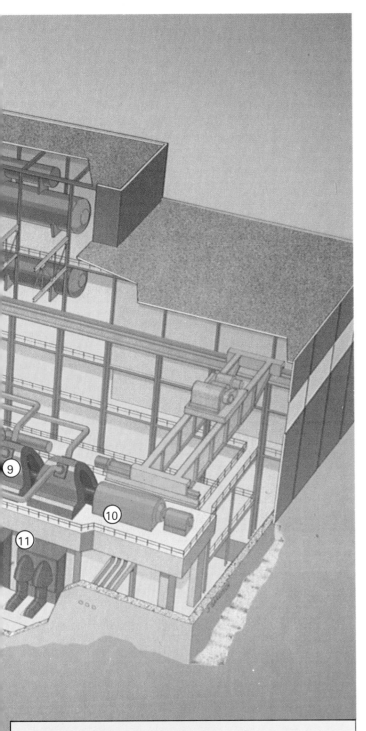

A nuclear power station has the same basic layout as those fuelled by coal, oil or gas. Water is heated in large boilers to produce steam. The steam is piped into turbines that turn generators, then it is cooled and condensed.

Although nuclear power stations do not produce any smoke or fumes, there is always a danger of radioactive gases leaking into the environment. They are usually sited in fairly remote areas. Within the power station itself, every precaution is taken against the accidental escape of radioactivity.

The nuclear reactor itself is housed inside a separate building with very thick concrete walls. The reactor's coolant system transfers heat to heat exchangers that produce steam. The heat exchangers are also located in the reactor building. They are fed by an outside supply of water, and the steam leaves through underground pipes. This cuts to a minimum the time that the water and steam are exposed to radiation from the core.

Safety systems

The greatest danger is that the chain reaction in the core will get out of control, causing the reactor to overheat. A number of safety systems are operated to prevent this. When the temperature of the core starts to rise, control rods are automatically lowered into it to slow down the chain reaction. If necessary, they can be used to shut down the core completely. A secondary safety system, located above the reactor, can be used to drench the core with cold water. Apart from two major accidents, at Three Mile Island in the United States and Chernobyl in Russia, the nuclear power industry has a very good safety record.

◄ The CANDU (Canadian Deuterium/Uranium) reactor was first used in 1971 at the Douglas Point power station in Ontario, Canada. It is widely considered to be one of the safest and most efficient, and has been exported to several countries including Argentina and India. The reactor building consists of a cylinder of prestressed concrete, lined with additional shielding and reinforcement. The emergency drenching system is located just below the domed roof. Cold water for the steam condensers in the turbine room is taken from the nearby river. Other nuclear power stations take cooling water from the sea.

Key	
1 Reactor building	7 Steam in underground pipes
2 Nuclear reactor	8 Storage for radioactive fuel
3 Heat exchangers	9 Turbines
4 Water storage for emergency drenching	10 Generator
5 Drenching sprays	11 Water intake for condensers
6 Control room	12 Pumphouse for cooling water
	13 Water outlet
	14 Stand-by diesel generator

Wind and water turbines

Wind and water turbines work on the same principles as steam turbines, but operate at much lower pressures. The amount of energy in running water depends on the vertical distance that the water has fallen. This distance is known as the head of water. The greater the head, the greater the energy provided by a given quantity of water. In places where the head of water is above 30 m, it is sufficient to produce a high-pressure water jet. Electricity can then be generated by using a simple impulse turbine, such as the Pelton wheel.

Most hydroelectric power stations use a type of turbine perfected by the American engineer James Francis in the mid-1800s. The Francis turbine is a reaction turbine. The angled blades are completely immersed in water, and are driven by the flow of the water through them, rather than by the impact of water against them. This enables the Francis turbine to operate with only a few metres of head, that is, by water with less pressure.

A Francis turbine consists of a single rotor, also called a runner, fitted with angled blades. The rotor is mounted within a spiral volute chamber. The turbine can be installed vertically or horizontally. In most cases, the rotor is mounted horizontally, and the electric generators are powered by a vertical shaft.

Where the head of water is too low for a reaction turbine, special low-head turbines can be used. One such design, the Kaplan turbine, uses a runner shaped like a ship's propeller. The Kaplan turbine can be fitted within a volute chamber, or can be placed directly into the flow of water.

Wind turbines have been developed from the traditional windmill, and many different designs are now in use. Most installations use horizontal-axis turbines with rotors shaped like an aeroplane propeller. Vertical-axis wind turbines, such as the Darreius turbine, are less common. Their advantage is that they can operate whatever the wind direction.

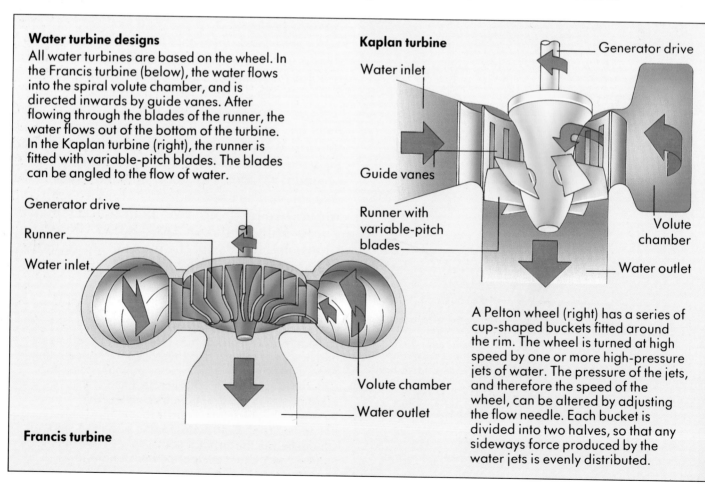

Water turbine designs
All water turbines are based on the wheel. In the Francis turbine (below), the water flows into the spiral volute chamber, and is directed inwards by guide vanes. After flowing through the blades of the runner, the water flows out of the bottom of the turbine. In the Kaplan turbine (right), the runner is fitted with variable-pitch blades. The blades can be angled to the flow of water.

Generator drive
Runner
Water inlet
Volute chamber
Water outlet

Francis turbine

Kaplan turbine
Generator drive
Water inlet
Guide vanes
Runner with variable-pitch blades
Volute chamber
Water outlet

A Pelton wheel (right) has a series of cup-shaped buckets fitted around the rim. The wheel is turned at high speed by one or more high-pressure jets of water. The pressure of the jets, and therefore the speed of the wheel, can be altered by adjusting the flow needle. Each bucket is divided into two halves, so that any sideways force produced by the water jets is evenly distributed.

▼ The generator room at a hydroelectric power station in the Snowy Mountains in Australia. The generators themselves are installed on the next level down, and the water turbines are on the next level below that. The power station forms part of a mammoth hydroelectric and irrigation scheme. The scheme involved diverting the Snowy River back on itself so that it can water arid regions.

Pelton wheel

Water jet

Generator drive

Buckets

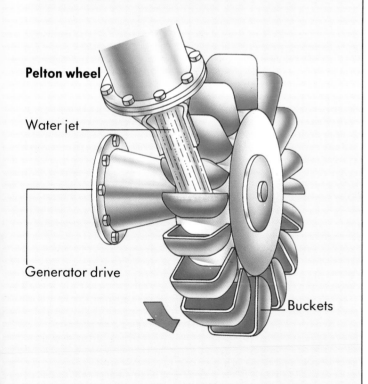

Rotor blades

Gearbox

Generator

Current flow

▲ A horizontal-axis wind turbine. Most designs use a three-bladed rotor, although some have only two. The main disadvantage of horizontal-axis turbines is that they can only operate when the wind is blowing straight at them.

Powerhouses

A power station produces electric current in massive generators. These are often called turbo-generators, because they are driven by steam or water turbines. A more correct name is turbo-alternators, because they produce alternating current.

In power stations that use fossil or nuclear fuels, the turbo-alternators are located in a separate powerhouse. Steam for the turbines is sent along insulated pipes from the reactor or boiler room. Turbo-alternators are designed to operate at a constant speed, and this is adjusted by varying the pressure of the steam entering the turbines. In the United States, turbo-alternators operate at 3,600 rpm to produce alternating current at 60 hertz (cycles per second). In Europe, they rotate at 3,000 rpm to produce current at 50 cycles.

The high speeds of operation mean that turbo-alternators also produce large amounts of unwanted heat. Modern designs are filled with hydrogen gas under pressure, which enables them to lose heat more efficiently. Water is circulated through the outer casing in order to carry the unwanted heat away.

Electricity is taken from each turbo-alternator by three cables in order to produce a three-phase supply. The cables run under the powerhouse floor to the bus room. In the bus room, electricity from the various turbo-alternators is brought together at a central point. The bus system usually consists of a network of heavy copper bars or cables, separated by large insulators.

Individual sections of the bus system are separated by circuit-breakers. If there is a fault in any part of the system, these automatically stop the flow of current, and isolate the faulty section. The flow of electricity can also be regulated by heavy-duty switchgear, operated from the power station's control room. Before electricity leaves the power station, it passes through a series of transformers, which produce the very high voltages required for long-distance transmission.

As well as the main turbo-alternators, most power stations also have smaller generators powered by gas turbines. These are used to produce additional electricity during periods when demand is especially high.

▲ The bus system and switchgear are protected against power surges by automatic circuit-breakers.

◀ Supervisors in the control room constantly monitor the output of each turbo-alternator. They route the electricity through the bus system to the main transformers.

▲ In the final stage of electricity production, the transformers step up the current to very high voltages.

▼ The powerhouse of a coal-fired power station. Steam for the lines of turbo-alternators is piped up from the boiler room below. Cables carrying electric current to the bus room are set into the floor. Overhead cranes are used to lift up the massive outer casings of the turbo-alternators when they need servicing and maintenance.

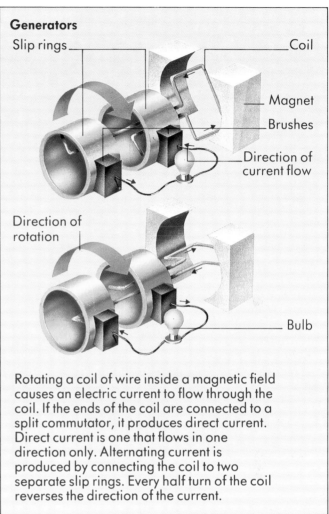

Generators

Slip rings — Coil

Magnet

Brushes

Direction of current flow

Direction of rotation

Bulb

Rotating a coil of wire inside a magnetic field causes an electric current to flow through the coil. If the ends of the coil are connected to a split commutator, it produces direct current. Direct current is one that flows in one direction only. Alternating current is produced by connecting the coil to two separate slip rings. Every half turn of the coil reverses the direction of the current.

Using electricity

Spot facts

• By 1900, more than two-and-a-half million light bulbs were being used to light London.

• In the United States, some power lines carry electricity at 765,000 volts.

• The power stations with the greatest output are hydroelectric. Currently the one with the most capacity is the plant at Grand Coulee in Washington State, USA, which has an output of 7.4 megawatts (million watts).

• The biggest ever power cut occurred in 1965 when the city of New York was blacked out for 13 hours.

► Coloured lights on the Christmas tree are now a familiar sight in many parts of the world. Traditionally, Christmas trees were lit with candles, which meant there was a constant risk of the tree catching fire. Electric lighting is much safer, but proper precautions must still be taken. Most Christmas tree lights operate on low voltages supplied by a separate transformer.

Electricity is our most useful form of energy; it is instantly available at the touch of a switch. During the last 100 years, electricity has completely transformed human society. Electricity provides basic heating and lighting, and also powers a wide range of machines. In the developed countries, many homes now contain over a dozen different electrical appliances.

Power stations produce high-voltage electricity, which is carried by power lines for hundreds of kilometres, to be used by factories and houses. Household appliances, however, use much lower voltages, and during the distribution of electricity, the voltage is progessively reduced. Inside the home, electricity is carried by a number of different circuits.

Transforming society

The single most important electrical device is the light bulb. Before its invention, the only sources of artificial lighting were oil and gas lamps. The first practical electric light bulbs were developed during the 1870s by Joseph Swan in Britain, and by Thomas Edison in the United States. Electric lighting had the advantage of being fairly cheap to install, and it was much safer than oil or gas. By 1900, electricity was in use throughout the world.

Other useful electrical appliances soon followed. By 1890, electrical engineers had perfected the electric fire, with long-lasting heating elements made from nickel-chromium alloy. Electric cookers and vacuum cleaners first went on sale in the 1890s, but were not widely used before the 1920s.

Since 1920, electricity has completely transformed daily life, and a wide variety of household appliances are now available. Electric irons, food mixers, washing machines, refrigerators, air-conditioning units and microwave ovens are now widely used.

Electricity also provided the energy for new forms of mass communication and popular entertainment. First radio, and then cinema and television, were made possible by the availability of electrical energy.

Industry too was transformed; electricity is now used in many basic processes, especially in the metal-working and chemical industries. More recently, electricity has allowed the development of electronic computers and the revolution in information technology.

◄ The Crystal Palace was built in London during the age of gas and oil lamps. In 1885 it was lit by electricity and provided a dramatic advertisement for the new energy source.

▼ The electric light bulb invented by Thomas Edison in 1879. The first light bulbs had a carbon filament made from cotton thread, which only lasted for about 40 hours. By 1900 there were longer-burning filaments made from metal wire. Modern bulbs use tungsten filaments.

Distribution

High-voltage power lines are used to carry large quantities of electricity over long distances. Energy is lost, mainly in the form of heat, as electricity travels down wires and cables. The use of very high voltages, typically around 225,000-275,000 volts, reduces this energy loss to a minimum. The most efficient and widely-used distribution method is by overhead power lines, supported by metal pylons. The atmosphere provides cooling, and most of the insulation. At points where the power line passes under the arm of a pylon, it is encased in a glass or ceramic insulator.

Underground high-voltage lines are much more expensive to install, and are normally used only over short distances. The power line itself is wrapped in several layers of plastic insulation and has a strong outer casing bound with steel wire.

When electricity is taken from power lines, the voltage is stepped down by banks of transformers at a substation. Most substations also contain circuit-breakers and switchgear.

Stepping down the voltage to the levels used by the customer usually requires a series of substations. Factories often use electricity at 10,000 volts or more, which is much too high for domestic use. In cities, a series of local substations provide electricity for distribution to houses and shops. Large office buildings usually take electricity at a higher voltage than they can use directly, and have a separate substation in the basement.

In isolated regions, a power station may be constructed to provide electricity directly to a town or large factory. In most cases, however, all the power stations in a country are linked into a national grid of power lines. This enables electricity to be switched to wherever it is needed. Output from power stations can also be matched more closely to demand, and electricity can even be traded between different national grids. This normally takes place by overhead power line. Britain, however, imports some of its electricity from France along an undersea cable.

▼ Electricity is generated as alternating current at a power station, where transformers step up its voltage to several hundred thousand volts. Power lines carried by tall pylons transmit the high-voltage current to local substations, where transformers step down its voltage. Factories may take the current at 10,000-30,000 volts. Homes and offices use electricity at mains voltage, which is usually about 240 volts.

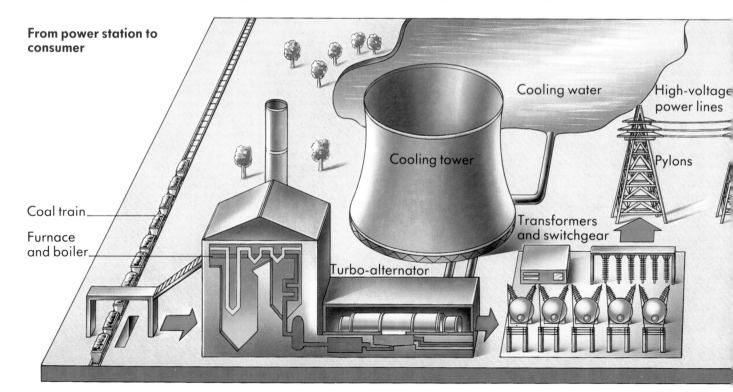

From power station to consumer

Coal train

Furnace and boiler

Cooling tower

Turbo-alternator

Cooling water

High-voltage power lines

Pylons

Transformers and switchgear

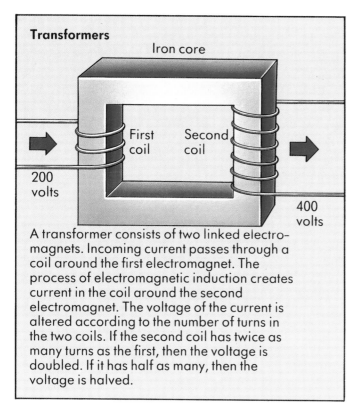

Transformers

Iron core

First coil

Second coil

200 volts

400 volts

A transformer consists of two linked electro-magnets. Incoming current passes through a coil around the first electromagnet. The process of electromagnetic induction creates current in the coil around the second electromagnet. The voltage of the current is altered according to the number of turns in the two coils. If the second coil has twice as many turns as the first, then the voltage is doubled. If it has half as many, then the voltage is halved.

▲ A pylon carrying high-voltage power lines, which consist of sets of three cables. They have an aluminium conductor and are reinforced with steel.

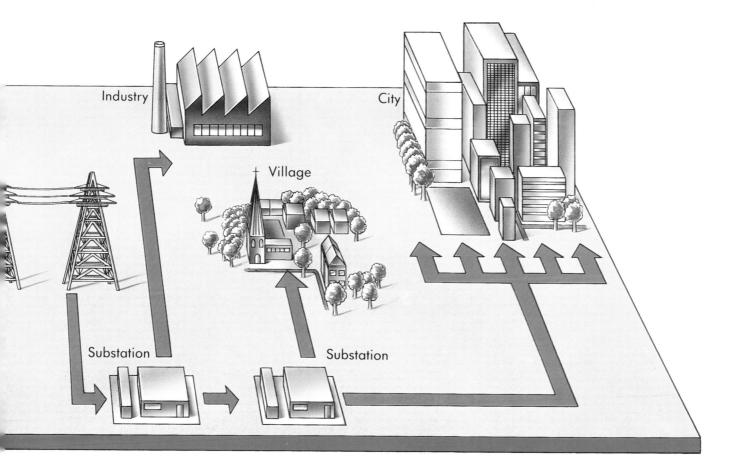

Industry

City

Village

Substation

Substation

Heating and lighting

Electricity gives off heat when it meets resistance to its flow. Useful quantities of heat can be produced by passing electricity through a coil of thin wire. A thin wire has a higher resistance to electricity than a thick wire. Winding the wire into a coil concentrates the heating effect into a small area. Coils used for heating are called electric elements, and are found in many household appliances. Maximum heat is produced when the element is glowing red hot.

An electric light bulb works by heating a coil of very thin wire, known as a filament, until it glows white hot. The first light bulbs contained a vacuum to prevent the filament from burning up. Modern bulbs are filled with an inert gas such as argon. A light bulb produces light across the whole range of the visible spectrum, but is very inefficient in its use of electricity. Only about 6 per cent of the energy is released as useful light; the rest is lost as heat.

Discharge lamps are much more efficient at turning electricity into light. A discharge lamp consists of a glass tube containing a gas or vapour that conducts electricity. Metal contacts at each end of the tube allow current to pass through the gas. The flow of electricity causes the gas atoms to agitate, and as a result they give off light.

Different gases produce different colours of light. Sodium vapour produces orange light, and mercury vapour produces blue light. Discharge lamps are widely used for street-lighting, but the colour of the light makes them unsuitable for use in the home or office.

Discharge lamps filled with neon gas can produce a wide variety of colours and are used in illuminated signs, often for advertising purposes. Other types of electric lamp are designed to produce infrared light, and are often used as bathroom heaters.

Fluorescent light tubes

Fluorescent tubes contain mercury vapour, and produce white light by means of a coating of phosphors on the inside of the tube. As well as visible blue light, mercury vapour also gives off a great deal of invisible ultraviolet light. The phosphors absorb this and give off a white light which is suitable for most purposes.

Fluorescent tubes use much less electricity than ordinary light bulbs and last much longer. Small fluorescent tubes are now being incorporated in "economy" light bulbs that can be substituted for the ordinary type.

▶ The city of Boston, USA, by night. Electric lighting has greatly extended the working day, and offices and other buildings can now operate around the clock. Outside, powerful electric floodlighting also enables open-air activities, such as football matches and other sporting events, to take place in the evening.

The fluorescent tube

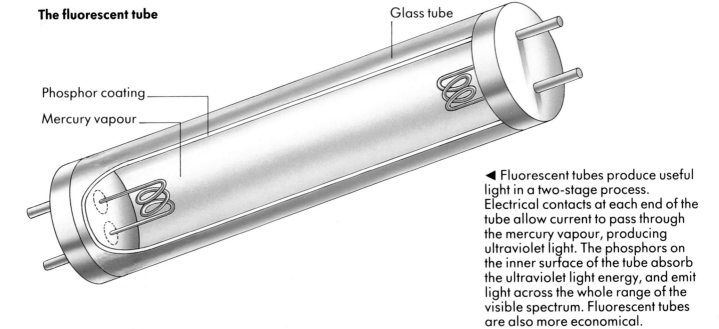

Glass tube

Phosphor coating

Mercury vapour

◀ Fluorescent tubes produce useful light in a two-stage process. Electrical contacts at each end of the tube allow current to pass through the mercury vapour, producing ultraviolet light. The phosphors on the inner surface of the tube absorb the ultraviolet light energy, and emit light across the whole range of the visible spectrum. Fluorescent tubes are also more economical.

Heating by radio

A microwave oven heats food by means of high-frequency radio waves, known as microwaves, which have a very short wavelength. The microwaves are produced by a device known as a cavity magnetron. As the microwaves pass through an item of food, they cause its molecules to vibrate, and this vibration produces heat. The cavity magnetron was developed during World War 2 in order to produce tightly-focused beams of microwaves for the first radar systems. Although the microwaves are said to form a continuous beam, the magnetron is in fact switching on and off about 10,000 times every second. The great advantage of microwave cookers is that they heat food extremely quickly.

Circuits and wiring

The electrical wiring inside a house consists of a number of ring circuits. The supply of electricity from outside, which is often called "the mains", is split into the different circuits at the fuse box. All the ring circuits use the same voltage as the mains, but are designed to carry different loads of electricity. Each circuit has a separate fuse, which melts if the circuit becomes overloaded. Some modern houses now use circuit-breakers instead of fuses.

Each floor of a house usually has two ring circuits: a power circuit and a lighting circuit. The power circuit supplies electricity to wall sockets, and is used for appliances which require large amounts of electricity. Appliances such as heaters, cookers and televisions must always be connected to the power circuit. Each socket is wired into a loop within the power circuit, and is controlled by a two-way switch.

The lighting circuit is normally built into the ceiling, and supplies electricity to individual light fittings. Light switches are usually wall-mounted and are connected to the lighting circuit by longer loops of wiring.

Dangers of electricity

Electricity is dangerous. The current of mains electricity is high enough to kill people. Do not attempt to investigate or experiment with any of the electrical fittings in your home.

Domestic wiring

▼ Electricity used in the home is often called mains electricity because it comes direct from the local substation. The fuse box is a safety device to isolate the different circuits from each other, and does not alter the voltage of the electricity in any way.

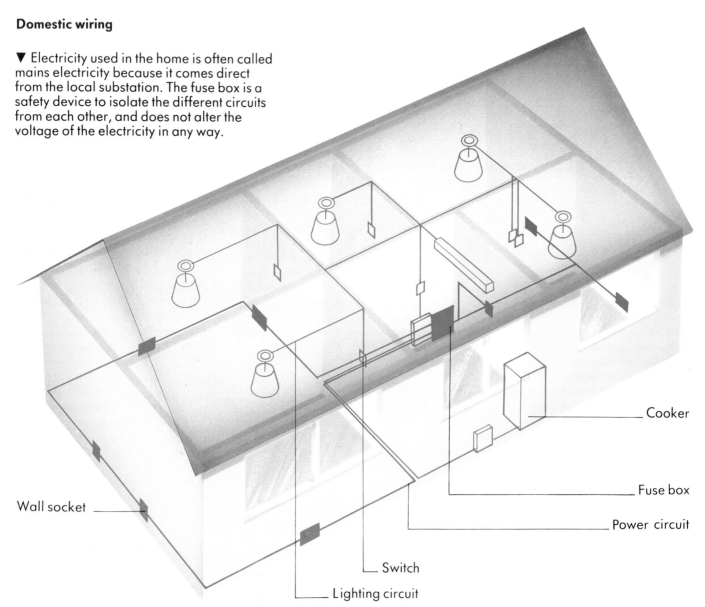

Cooker

Fuse box

Power circuit

Wall socket

Switch

Lighting circuit

Motors

Electric motor

An electric motor consists of a coil of wire attached to a shaft, and surrounded by a simple magnet. When electricity flows through the coil, the effect of the magnetic field causes the coil to rotate. The supply of electric current to the coil is controlled by a split-ring commutator and brushes.

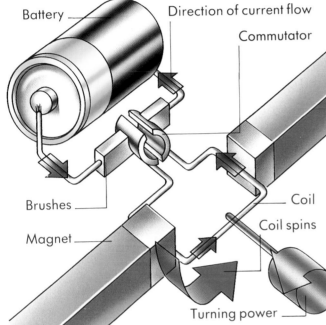

Battery
Direction of current flow
Commutator
Brushes
Coil
Magnet
Coil spins
Turning power

Apart from heating elements and lights, the most useful electrical devices are electric motors. Motors of many different sizes are built into many household appliances. Some use the motor's power directly, some indirectly.

Electric drills and food blenders make direct use of the motor's rotating shaft. Vacuum cleaners and hairdriers use motors to turn a fan which creates a flow of air. This can be used to pull air into the appliance, or blow air out.

Water pumps driven by electric motors are found in washing machines, shower units and central heating systems. Refrigerators also use electric motors; in these appliances they circulate special coolant liquids.

▼ A hairdrier uses electricity in two different ways. A small electric motor drives a fan, and heat is produced by electric elements in the form of wire coils. As a safety precaution, the drier is fitted with a heat-sensitive bimetallic switch. If the drier overheats, the switch opens, and breaks the electrical circuit.

Electric hairdrier

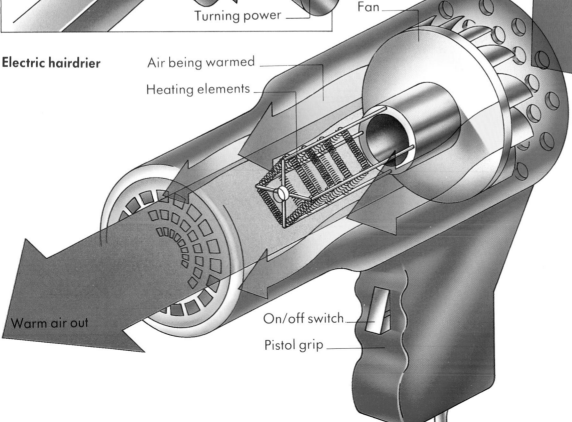

Air being warmed
Heating elements
Fan
Cool air drawn in
Warm air out
On/off switch
Pistol grip

Engines

Spot facts

• A mass of steam has a volume 1,800 times greater than that of the same mass of water.

• China is the only country that is still building steam locomotives.

• One of the most powerful petrol engines ever built had 28 cylinders arranged in four circles of seven, and was used in American bomber aircraft during World War 2.

• Diesel engines of over 3,000 horsepower are used to power many modern locomotives.

▶ A scale-model traction engine pulls a young visitor at a steam-power exhibition. Full-size traction engines were used in farming. As well as providing motive power, they were also used to drive agricultural machinery, by means of moving belts. Traction engines were also used as the first steamrollers, hence the name.

Engines are devices that burn fuel and provide mechanical power. Steam engines and petrol engines both produce power through the up-and-down motion of a piston inside a cylinder. Engines that operate in this manner are called reciprocating engines.

Steam engines burn fuel separately to produce steam in a boiler. They are therefore known as external combustion engines. Petrol engines burn fuel inside the engine itself, and are internal combustion engines.

The age of steam

If steam is condensed within a closed container, a vacuum is created. The first practical steam-powered device, invented by the Englishman Thomas Savery in 1698, was a vacuum pump.

In 1712, another Englishman, Thomas Newcomen, built the first true steam engine that used a piston within a vertical cylinder. Steam entering at the bottom of the cylinder forced the piston up. The steam was then condensed by a spray of cold water. The resulting vacuum allowed the piston to fall, ready to be driven up again by more steam. A connecting rod transferred the up-and-down motion to a hinged beam, hence the name beam engine.

Newcomen's engines produced a steady pumping action, and were widely used in coal mines. But they were slow and inefficient.

Between 1769 and 1790, the Scottish inventor James Watt made a number of improvements, making the steam engine flexible and efficient. Watt's first improvement was the separate steam condenser, which allowed the cylinder to remain hot. He also invented double-action, in which both the up stroke and the down stroke were steam-powered. His final improvement was to connect the piston to an offset flywheel, which converted the up-and-down motion into rotation.

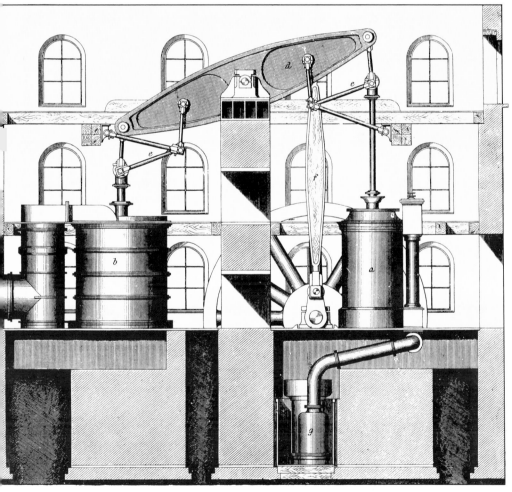

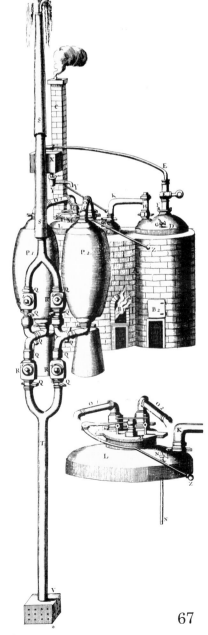

▲ A contemporary print showing an industrial steam engine of the mid-1800s. The right-hand arm of the beam (d) was pulled down when steam expanded against a piston in the steam cylinder (a) and was condensed in a condenser (g).

▶ Savery's steam pump used two boilers side by side and was very unsafe. Water was alternately boiled and condensed in each of the two boilers. The vacuum that was produced drew water up a pipe and through a one-way valve.

Steam engines

During the 1800s, the steam engine was further transformed. Vertical cylinders were replaced by horizontal ones, and better engineering techniques allowed high-pressure steam to be used with safety. This meant improved boilers, with steam being heated to progressively higher temperatures in a series of tubes within tubes. As a result, steam engines became smaller and produced more power.

Some designs recycled steam through a series of expansion chambers and condensers, thus making maximum use of steam's heat energy. Such engines tended to be both large and heavy, and were mainly used in factories.

Other designs abandoned bulky condensers and discharged steam straight into the atmosphere once it had been used. These smaller and lighter engines were mainly used in steam locomotives.

A steam locomotive is a self-contained unit that produces enough power to pull heavy loads along railway lines at high speeds. Fuel, usually coal, is burned in the firebox to produce heat for the boiler. Steam locomotives normally have two cylinders, one on each side. The movement of the pistons in the cylinders is transferred to the driving wheels through a series of connecting rods and cranks.

Steam locomotive

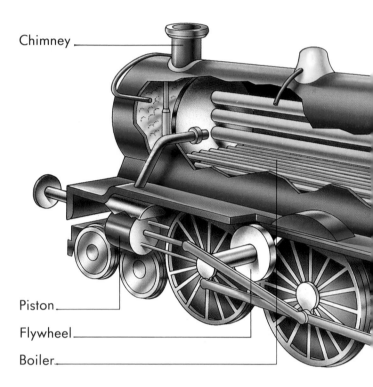

Chimney

Piston

Flywheel

Boiler

in the 1880s of the petrol-driven motor car. Heavy steam-powered traction engines, however, found widespread use on farms and for road-construction work for many years.

Stationary steam engines of all sizes were used throughout the 1800s to produce power in factories throughout Europe and North America. But by 1900 large steam engines were being replaced by the more efficient steam turbines, and smaller ones by internal combustion engines, such as gas engines.

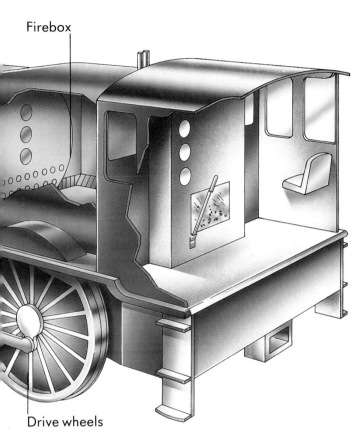

Firebox

Drive wheels

▲ In a locomotive hot gases from the firebox are drawn through the boiler tubes and boil the water. The steam is fed into the cylinders to drive the pistons, which are connected by rods to the drive wheels.

◄ A powerful steam traction engine, of the type used on farms from the late 1800s. They were used to haul ploughs and power machinery such as threshers.

The first steam locomotive, designed by the English engineer Richard Trevithick, appeared in 1804. By then steam power had also been applied to transport on water. A paddle steamer called the *Charlotte Dundas* had been operating in Scotland since 1801. And by 1819 steam boats were crossing the Atlantic. Later, steam carriages and cars enjoyed a brief popularity, but they became obselete with the development

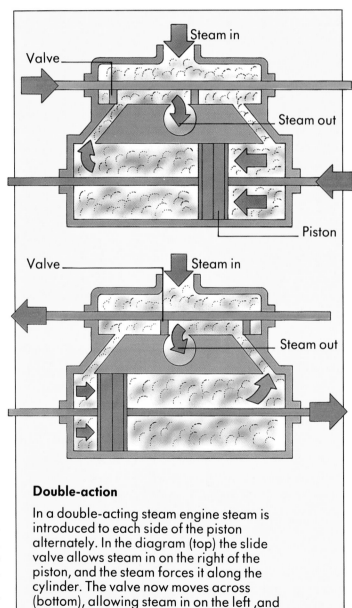

Steam in
Valve
Steam out
Piston
Valve
Steam in
Steam out

Double-action

In a double-acting steam engine steam is introduced to each side of the piston alternately. In the diagram (top) the slide valve allows steam in on the right of the piston, and the steam forces it along the cylinder. The valve now moves across (bottom), allowing steam in on the left ,and the spent steam to escape. James Watt introduced double-action in 1781.

Petrol engine

The petrol engine is an internal combustion engine. Fuel is burned inside a cylinder to provide the energy to drive a piston. Only one in four of the piston strokes produces power, and for this reason petrol engines are known as four-stroke engines.

The four-stroke engine was invented in 1876 by the German engineer N.A. Otto. The Otto engine burned a mixture of coal gas and air. In 1885, another German, Gottleib Daimler, invented the four-stroke petrol engine. Many improvements were made by other engineers, and by 1900, the petrol engine had acquired all the main features that it has today.

Petrol engines burn a mixture of air and petrol vapour in a ratio of about 14 parts air to one part petrol. The mixture is compressed by a piston and is then ignited by an electric spark. The combustion, or explosion, pushes the piston down smoothly. The downward motion is transformed into rotation by the crankshaft.

Because of the four-stroke cycle, petrol engines operate most efficiently when they have four cylinders operating in sequence. One of the cylinders is always on the power stroke, which means that the crankshaft is under constant drive. The rotating crankshaft is the source of all the useful power produced by the engine. Most of this power is used to perform work, for example propelling a car. Some rotary motion is used to operate engine components.

Petrol engines are mainly used in cars and motorcycles. Many car engines have four cylinders, and larger engines often have multiples of four. The cylinders can be arranged in a line, or in a V-shape. Engines are often described as V4, V6, V12 and so on. Motorcycle engines often have only one or two cylinders.

Petrol engines are capable of sustained operation at high speed. When producing maximum power, the pistons may travel up and down the cylinders 175 times every second.

◄ A high-performance rally car cornering at speed. Many high-performance engines use superchargers or turbochargers to compress the air, so that a greater quantity of fuel can be burned at each stroke, thus producing more power.

Four strokes

The first stroke in the cycle (intake stroke) draws the petrol/air mixture into the cylinder. The second stroke (compression stroke) compresses the mixture, which is ignited when the piston is at the very top of the up stroke. During the third stroke (power stroke) the piston is driven down by the rapidly-expanding gases produced by combustion. The fourth stroke (exhaust stroke) forces exhaust fumes through the outlet valve, and the cylinder is ready to repeat the cycle. Only the power stroke drives the piston. During the other three strokes, the pistons are driven by the rotation of the crankshaft.

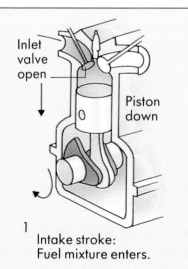

Inlet valve open

Piston down

1 Intake stroke: Fuel mixture enters.

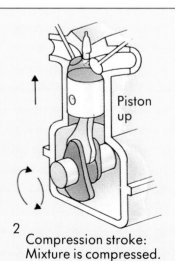

Piston up

2 Compression stroke: Mixture is compressed.

The petrol engine

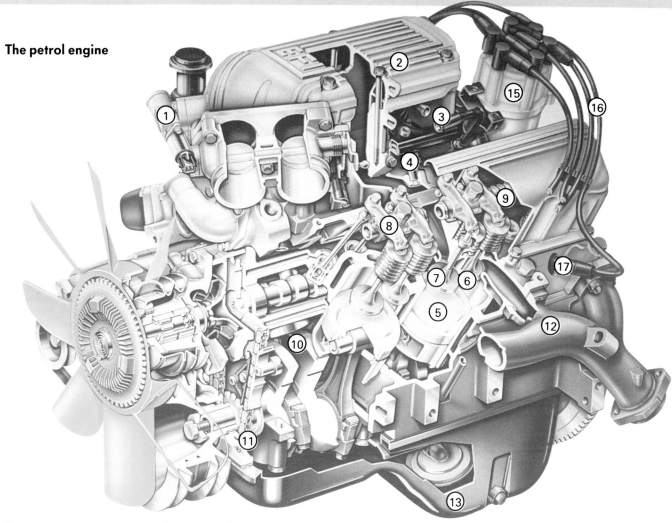

Key
1 Air intakes
2 Air chamber
3 Fuel rail
4 Fuel injector
5 Piston
6 Outlet valve
7 Inlet valve
8 Valve rocker
9 Camshaft
10 Crankshaft
11 Chain drive for camshaft
12 Exhaust manifold
13 Sump
14 Flywheel
15 Distributor
16 High-voltage lead
17 Spark plug connector

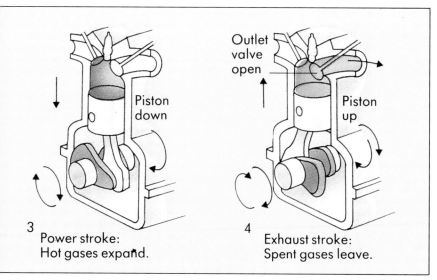

3 Power stroke:
Hot gases expand.

4 Exhaust stroke:
Spent gases leave.

▲ Cutaway drawing of a modern V6 petrol engine, typical of those used in medium-sized family cars. The cylinders are arranged in two rows of three, set at about 120° to each other. Air enters the intakes and passes through the air chamber to the inlet valves. Petrol is pumped along the fuel rail and is injected into the air flow and vaporizes. After combustion, exhaust fumes pass through the outlet valves and into the exhaust manifold. Most of the power produced by the engine is transmitted by the crankshaft to the flywheel at the back of the engine. At the front of the engine, the crankshaft also drives both the cooling fan and the overhead camshaft. The camshaft operates the valves by means of a series of rods. The rods are connected to spring-loaded rockers.

71

Engine systems

As well as fuel and air, a petrol engine also requires electricity, oil and water. These are supplied by different systems.

The fuel system pumps liquid petrol from the fuel tank and distributes it in measured quantities into the air flow to the cylinders. Most engines regulate the flow of petrol with mechanical devices known as carburettors. Some modern engines now use electronically-controlled fuel injectors. The engine's breathing system draws in air through a filter, and channels the air flow to the intake valves.

The exhaust system takes hot gases from the cylinders through the exhaust manifold and out of the engine. In most cases, a silencer is fitted to the end of the exhaust system. Some countries also require that exhaust gases pass through a catalytic converter to reduce the amount of pollution emitted into the atmosphere.

The engine's lubrication system ensures that all moving parts are coated with a thin film of oil. This reduces friction. The main oil reservoir is the sump, which is located at the base of the engine beneath the crankshaft.

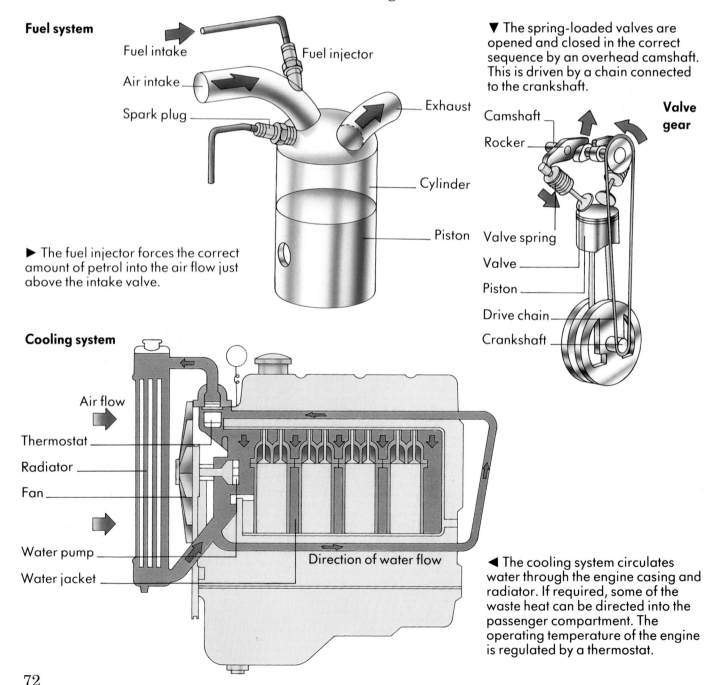

Fuel system

Fuel intake

Air intake

Spark plug

Fuel injector

Exhaust

Cylinder

Piston

▶ The fuel injector forces the correct amount of petrol into the air flow just above the intake valve.

▼ The spring-loaded valves are opened and closed in the correct sequence by an overhead camshaft. This is driven by a chain connected to the crankshaft.

Valve gear

Camshaft

Rocker

Valve spring

Valve

Piston

Drive chain

Crankshaft

Cooling system

Air flow

Thermostat

Radiator

Fan

Water pump

Water jacket

Direction of water flow

◀ The cooling system circulates water through the engine casing and radiator. If required, some of the waste heat can be directed into the passenger compartment. The operating temperature of the engine is regulated by a thermostat.

72

▶ All the moving parts in an engine receive constant lubrication. A pump draws oil from the sump, or reservoir, and forces it under pressure through channels to the engine bearings on the crankshaft and to the camshaft. Other areas get splashed with oil. Despite lubrication, some engine wear takes place. For this reason the system incorporates a filter to remove solid fragments before they do damage.

▼ A spark plug creates an electric spark across a tiny gap between two metal contacts, or electrodes. The inner electrode is connected with the plug lead and receives high- voltage current from the ignition coil via the distributor. The outer, or earth electrode is connected to the plug casing. The two electrodes are insulated from each other by a thick ceramic layer.

Spark plug

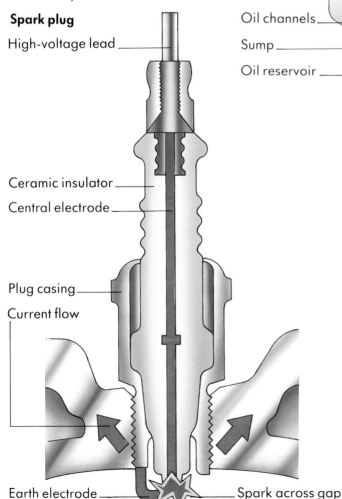

High-voltage lead

Ceramic insulator
Central electrode

Plug casing
Current flow

Earth electrode — Spark across gap

Lubrication system

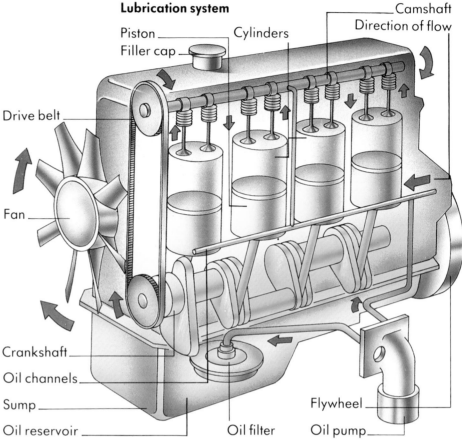

Piston
Filler cap
Cylinders
Camshaft
Direction of flow
Drive belt
Fan
Crankshaft
Oil channels
Sump
Oil reservoir
Oil filter
Flywheel
Oil pump

Some engines are designed to be air-cooled. They radiate away unwanted heat directly into the atmosphere. Air-cooled engines, such as many motorcycle engines, usually have a series of cooling fins around the cylinder casing.

Most petrol engines use both air and water to carry heat away from the cylinders. Water circulates through the engine and radiator inside a sealed system. Heat from the engine heats the water, which is then cooled inside the radiator. A stream of air is drawn through the radiator by a fan mounted at the front of the engine. The fan is usually turned by a belt attached to the crankshaft.

The electrical system is based around a low-voltage (12 v) battery. The battery supplies current to the ignition coil, which produces the high voltage needed by the spark plugs. High-voltage current is supplied to each spark plug in turn by the distributor, which is a kind of high-speed rotary switch. The battery is kept under charge by a generator driven by the crankshaft. The electrical system may also be used to drive the fuel, oil and water pumps.

Other engines

Other types of piston engine share the same basic design as the petrol engine, but differ slightly in the details of their operation. Some, such as the diesel engine, are designed to use a different type of oil-based fuel.

The diesel engine was invented by the German engineer Rudolf Diesel in 1893. Like the petrol engine, the diesel engine is a four-stroke engine. It differs, however, because it has no separate ignition system. The diesel engine is self-igniting, and uses the principle that air heats up when it is compressed. The action of the piston in a diesel engine compresses the air taken in so much that it becomes hot enough to ignite the fuel.

Diesel engines are more efficient than petrol engines, but tend to be quite heavy. They are mainly used in large vehicles, such as trucks, buses, agricultural vehicles and railway locomotives. Small diesel engines are now fitted in some cars. Diesel engines are also widely used to power electrical generators.

▲ Trucks powered by large diesel engines are used in Australia to pull road-trains, consisting of several trailers. Once the road-train gets up to its cruising speed, the diesel engine performs very efficiently. At slow speeds, however, this form of transportation is extremely expensive because of its high level of fuel consumption.

◄ Motorcycle sport is popular in many countries, especially with the young. Two-stroke engines have a much simpler design and fewer moving parts than four-stroke engines. Though not as powerful, they are generally more reliable. This makes them ideal for young riders who like to do their own servicing and maintenance.

The two-stroke engine is a petrol engine that produces power every other stroke. Two-stroke engines have no separate lubrication system, and lubricating oil has to be mixed into the petrol. Combustion is less efficient, and they produce a great deal of exhaust fumes.

Some petrol engine designs do not use a piston within a cylinder. Rotary engines, such as the Wankel engine, use a flat, three-sided rotor that turns around a central shaft. It revolves in a figure-of-eight shaped chamber. As it revolves, it creates spaces in which the four stages in the four-stroke engine cycle take place. These four stages are intake, compression, power and exhaust.

The basic petrol engine can also be adapted to run on other fuels. During World War 2 many cars and lorries ran on gas. More recently, Brazil has tried out alcohol-powered cars.

Two-stroke engines

The operation of the engine ports (valves) is controlled by the piston. The fuel mixture of petrol, oil and air enters the inlet port and then transfers to the cylinder. The up stroke of the piston closes the transfer port; the fuel mixture is compressed and then ignited by a spark plug. During the down stroke (the power stroke), exhaust gases leave through the exhaust port, and fresh mixture flows in through the transfer port.

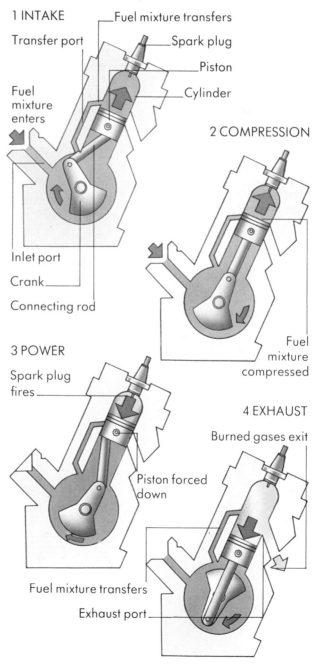

1 INTAKE
Transfer port
Fuel mixture transfers
Spark plug
Piston
Cylinder
Fuel mixture enters
Inlet port
Crank
Connecting rod

2 COMPRESSION
Fuel mixture compressed

3 POWER
Spark plug fires
Piston forced down

4 EXHAUST
Burned gases exit
Fuel mixture transfers
Exhaust port

Jets and rockets

• In 1947, the US pilot, Charles "Chuck" Yeager in the rocket-powered XS-1 became the first person to fly faster than the speed of sound.

• Modern jet engines are about 40 times more powerful than the biggest petrol engines.

• The Space Shuttle is the most complex machine ever built. It contains 49 rocket engines, 23 radar and radio antennae, and five computers.

• Russia's Energia rocket is currently the world's most powerful rocket, with a lift-off thrust of nearly 3,500 tonnes.

▶ A Harrier "jump jet" uses swivelling nozzles to direct the exhaust gases from its jet engine downwards as it takes off and lands vertically. It swivels the nozzles through 90° to produce horizontal thrust so that it can fly normally.

Jets and rockets produce power in the form of direct thrust. They burn fuel and concentrate a stream of exhaust gas through a nozzle at the back of the engine. This backwards flow of hot gas causes the engine to move forwards by reaction, and for this reason jets and rockets are called reaction engines.

Jet engines breathe air, and are now used in all types of aircraft and helicopters. Rockets carry their own oxygen supply and can produce far more power than jet engines, but can only burn for a fairly short period of time. During the period of combustion, rockets can achieve very high velocities: fast enough to escape the pull of Earth's gravity and travel into space.

The jet age

ntil the 1950s, most aircraft were propeller-driven. The rotary motion needed to turn the propellers was provided by high-performance petrol engines. The most powerful engines ould propel a single-seater aircraft up to about 60 km/h, but no faster. Higher speeds required different kind of engine, and by the 1930s many aircraft engineers were thinking about et propulsion.

The invention of the jet engine is generally redited to Frank Whittle, an officer in the British Royal Air Force. Whittle patented the asic design of the jet engine in 1930, but ndependent research was also taking place in ther countries. The first jet-powered flight was made by an experimental German aircraft, the Heinkel He-178, in August 1939.

Whittle's design was for what we today call a turbojet. Kerosene (paraffin) fuel is burned nside a stream of compressed air. The expanding exhaust gases drive a turbine before leaving the engine and providing thrust. Jet engines are also known as gas-turbine engines.

After World War 2, Whittle's engine was used in the Gloster Meteor, the first jet aircraft to enter regular service. The Meteor, which first flew in 1941, was a twin-engined jet fighter, with one engine set in each wing. The first jet-propelled passenger aircraft was the De Havilland Comet, which made its maiden flight in 1949, and entered service in 1952. Other passenger jets, including the Boeing 707, soon followed. By the 1960s, jets had largely replaced petrol-engined aircraft.

The main advantage of the jet engine is that it produces far more power than a petrol engine of the same size and weight. Jet engines can therefore propel larger aircraft at higher speeds over greater distances. Another advantage is they are mechanically simpler, producing rotary motion directly without the need for connecting rods and cranks. They also run more smoothly, with less vibration. And whereas the efficiency of propellers decreases with altitude, that of jets increases. Many aircraft now use turboprop and turbofan engines.

▼ Frank Whittle, pictured in 1944, the year Britain's Gloster Meteor jet fighter went into service with the RAF. It had made its first flight in 1941, two years after the first jet plane, the He-178.

▲ A modern turbofan engine, of the type used to power most of today's airliners. It differs from the turbojet of the Whittle design by having a huge fan in front. The fan not only directs air through the engine, it also directs it around the engine. For this reason it is sometimes called a by-pass turbojet.

Jet engines

The simplest form of jet engine is the ramjet, which has no compressor or turbine. A ramjet relies on the speed of its own motion through the atmosphere to compress the air entering at the front of the engine. Fuel is burned at the centre of the engine, and the expanding gases are directed through a nozzle at the rear.

The main disadvantage of the ramjet is that it cannot operate when the engine is not moving. Ramjets are therefore only used in missiles that are launched by a rocket motor. Once the missile is moving, the ramjet takes over.

A turbojet uses a rotary compressor at the front of the engine to provide high-pressure air to the combustion chamber. The exhaust gases pass through a turbine before leaving through the exhaust nozzle. The turbine provides just enough power to drive the compressor by means of a central shaft.

The power output of a turbojet can be boosted for short periods by afterburning. Additional fuel is burned in a second combustion chamber which is located between the turbine and the exhaust nozzle.

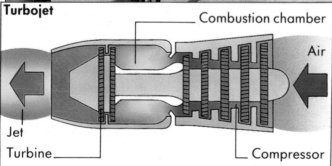

Turbojet

Combustion chamber

Air

Jet

Turbine

Compressor

▲ Turbojets are now mainly used in high-performance aircraft such as this two-seater Alphajet. The air intakes are situated on either side of the fuselage just below the cockpit canopy.

◀ The turbojet consists of a central combustion chamber with a compressor in front and a turbine behind. Modern engines use a multi-stage compressor to force air into the combustion chamber.

In the 1980s the United States began to phase out its ordinary, expendable rockets, like the Delta and Titan. It began to switch most of its satellite launchings to the reusable Space Shuttle. The Space Shuttle hardware is made of three main units. A winged orbiter carries the crew and payload (cargo). An external tank carries fuel for the orbiter's three liquid-fuelled main engines. Two solid-fuelled boosters, attached to the external tank, provide additional thrust at lift-off. The Space Shuttle can carry up to 20 tonnes of payload into low Earth orbit. After a mission the orbiter descends from space and lands on a runway like a normal aircraft.

The Space Shuttle has not proved as cheap or as reliable as had been hoped. And, following the *Challenger* disaster in 1986, the United States started using expendable rockets once again for routine satellite launchings. The Shuttle is now used mainly to support scientific missions. For example, it is used to carry the space laboratory, *Spacelab*, and to launch probes to the planets.

The European Space Agency (ESA) has also developed a range of powerful launching rockets, called Ariane. They launch satellites for international customers on a commercial basis. Ariane 3 is a three-stage launch vehicle that can place up to three satellites into geostationary orbit, 35,900 km high. ESA launches the Arianes from its launch site at Kourou, French Guiana, in South America.

In the Far East Japan, China and India have also developed launching rockets. Japan has a thriving space programme, and now launches space probes as well as satellites. In 1986 two of its probes encountered Halley's comet and returned much new information.

▼ A Russian rocket being delivered to the launch pad at the Baikonur Cosmodrome in Central Asia. Hydraulic rams will lift it into the vertical position alongside the launch gantry. The Russians have more experience at launching rockets than any other country. In Russia, a rocket can be set up and prepared for launch is about two days, compared with about 100 days in the United States.

Working with metals

Spot facts

- *Gold can be hammered into a sheet so thin that it becomes transparent. Just one kilogram could be beaten into a sheet 1,000 square metres in area.*

- *One kilogram of copper can be drawn into more than 1,000 km of fine wire without breaking.*

- *More than 10 million rivets were used to join the steel plates that form the hull of the transatlantic liner Queen Mary, now out of service.*

- *Some machine tools can cut with an accuracy of plus or minus 0.0025 mm.*

Most of the metals we use are produced by smelting ores at high temperatures in furnaces. They leave the furnace in a molten state. They then have to be processed into finished products. The shaping process selected depends on the metal concerned and what it is to be used for. A metal may be shaped when it is molten, when red-hot or when cold. It may be moulded, rolled, hammered, squeezed or welded. Afterwards it may be turned, ground, drilled or milled to very precise dimensions by machine tools. Precision machining holds the key to most manufacturing processes.

▶ White-hot steel pours into a travelling ladle from an electric-arc furnace. Next it will be cast into moulds, where it will solidify into ingots. These will then go to other machines for final shaping.

Casting

The technique of shaping metals by casting has been practised for at least 6,000 years. Casting is a process in which molten metal is poured into shaped moulds and allowed to cool. As it cools, it sets, or becomes solid, and takes the shape of the mould. Copper and bronze were the first metals shaped by casting because they could be melted in the early furnaces. Bronze is still widely used for casting, to make such things as statues and ships' propellers. The most common casting metal, however, is iron. Machinery bodies such as engine blocks are cast in iron, because cast iron is hard and rigid.

Sand casting

Casting takes place in a foundry. The most common method is sand casting. A model of the object to be made is placed in a box and a moist sand and clay mixture is packed tightly around it. A cavity of the required shape remains when the model is removed. Usually the mould is made in two halves to allow the model to be removed. Two holes are made in the top of the mould. The metal is poured in through one (called the runner), while the other (the riser) allows the air to escape from inside. The mould is broken up to release the casting when cool.

In a variation of this process, the model is made in wax. After the mould has been made, it is heated and the wax is poured out. Molten metal is then poured in. This method is called investment casting or the lost-wax (*cire perdue*) process. It is often used by artists, and to make precision castings for engines.

Diecasting

This is a method of casting that takes place in a metal mould, or die, which can be used over and over again. It is sometimes called permanent-mould casting. In gravity diecasting, molten metal is poured into a mould. It is suitable for producing simple shapes such as pipes.

More intricate shapes can be produced by injecting molten metal into a water-cooled mould under pressure. This method is widely used to make parts for machines and appliances. Alloys containing zinc, tin, aluminium and magnesium are favoured because they have a low melting point. Pressure diecasting is very suitable for mass-production.

▲ Workers handling a still red-hot casting of a railway wagon wheel, which has just been removed from its mould. It is cast in steel.

Sand casting

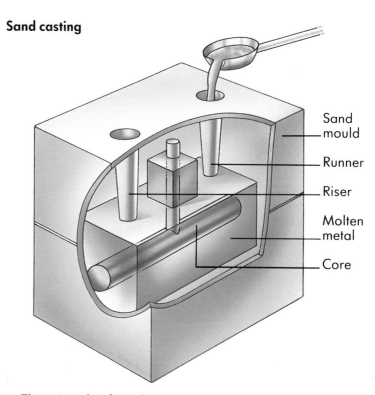

Sand mould
Runner
Riser
Molten metal
Core

▲ The principle of sand casting. Molten metal is poured into a hollow mould, which has the shape of the object to be made. To produce a hollow object, a core must be included. The metal is poured into the mould until it fills both the runner and the riser.

Rolling

Rolling is a process in which metal is passed between heavy rotating rollers, rather like clothes through an old-fashioned mangle. In a rolling mill the metal is passed through a succession of rollers, each pair being slightly closer together than those before. In this way the metal is squeezed thinner and thinner. Mostly, metal is rolled when it is red-hot. In this state it can "flow" more easily.

Much of the metal produced in furnaces is first cast into ingots. Usually these are then reduced to a more convenient size by rolling. The metal emerges as a flat slab, which may then go for shaping by another method, such as forging. Or it may be rolled further, for example, in a continuous strip mill. This produces coils of thin sheet, or strip. The slab goes slowly into the first set of rollers, but comes from the last set travelling at a speed of up to 100 km/h. The hot-rolled strip is then usually rolled again when cold. Cold rolling improves the surface finish and increases the hardness.

▲ Steel strip coming off a cold-rolling mill. During cold rolling the metal becomes brittle, which could cause it to crack in use. It therefore undergoes a heat treatment called annealing to bring it back to a reasonable condition. It is first heated and then it is allowed to cool slowly.

◄ High above the city streets, workers piece together a frame of iron girders that will support a new tower block. The girders are shaped with a typical H cross-section in rolling mills, using rollers with grooves cut in them.

Forging and pressing

Forging shapes metal by hammering. It is the oldest shaping method. In early times metals were forged by hand in much the same way that a traditional blacksmith does today. But in industry today forging is done by machine.

In a drop forge, the hammering action is produced by a falling hammer, or ram. The ram is raised again by air or steam pressure. Air or steam pressure may also be used to help accelerate the ram downwards to deliver an even more powerful blow. The ram shapes the metal by forcing it into a mould, or die.

Usually the ram carries the upper part of the die, while the lower part is mounted on the forge bed. The metal blank, usually hot, is placed on the lower die, and the ram is released. The metal is forced into shape as the two halves of the die come together. Stamping is a kind of small-scale drop-forging process used to make coins and medals.

On a forging press metal is forced into shape, not by a hammer blow, but by a gradual squeezing action. The press works by hydraulic (liquid) pressure. Some presses can exert a force of up to 50,000 tonnes. They are used, for example, to shape massive red-hot steel ingots. Smaller hydraulic presses are used to shape car body panels from cold steel sheet.

▲ A blacksmith practises the traditional craft of forging. A strip of metal is heated up in the forge fire, and then hammered into shape on an anvil. Metal is cooled by plunging it into cold water.

◄ Forging the rotor shaft of a turbine on a massive forging press. The shaft is beginning to take shape. It started off as an ingot casting, which was reheated until it was red hot. Then it was placed on the press and slowly squeezed into shape under a pressure of thousands of tonnes exerted by the hydraulic ram. Later, it will be machined on a lathe to bring it to the dimensions required in a rotor shaft.

Joining metals

Riveting

Many metal objects are so large or so complicated that they cannot be produced in one piece, but must be built up little by little. A ship's hull is an example. Until about the middle of the century, most hulls were built of steel plates joined together by rivets. A rivet is a metal plug with a rounded head at one end.

In riveting, holes are drilled in overlapping plates. Rivets are inserted through them and hammered to form a second head. The metal plates are then sandwiched tightly together. Riveting is no longer much used for producing ships' hulls, although it is still widely used elsewhere in shipbuilding. It is also used in aircraft construction for building the airframe and the outer "skin" of the fuselage and wings.

Welding

Most ships' hulls these days are constructed of steel plates that are welded together. The plates are joined edge to edge, with no overlap, which saves weight and materials.

In welding, the edges of the metal pieces to be joined are brought into contact and heated until they begin to melt and merge, or fuse together.

▶ A welder joins together two lengths of pipeline for the North Sea oilfields. The method is electric welding, using a special circular electrode. The welder wears thick protective clothing and a head mask as protection against the shower of sparks and glare which the welding process generates.

▼ These are the two main methods of producing a strong joint between pieces of metal. In riveting, headed rivets clamp overlapping metal plates together. In welding, joints are produced when touching pieces of metal melt and fuse together. Welded joints can be of different types, which include butt, lap, fillet and spot welds.

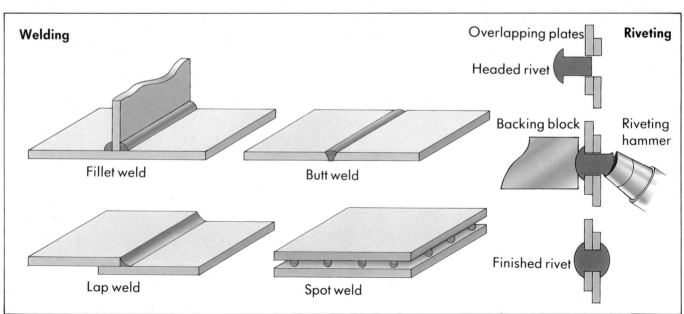

Welding

Fillet weld

Butt weld

Lap weld

Spot weld

Overlapping plates   **Riveting**

Headed rivet

Backing block   Riveting hammer

Finished rivet

Extra molten metal is often added from a so-called filler rod. When the metal in the joint cools, it forms a continuous structure linking the two pieces. The result is a strong joint.

There are three main methods of welding – gas, arc and resistance welding. In gas welding the heat to melt the metal pieces and filler rod is provided by an oxyacetylene torch. The torch is so called because it burns acetylene gas with oxygen, at a temperature approaching 3,000°C. Arc welding, on the other hand, uses an electric arc – a kind of continuous spark – to produce a high temperature. The arc is struck by passing a heavy electric current between the metal to be joined and an electrode held by the welder.

In resistance welding two electrodes carrying heavy current pinch together two overlapping plates. The resistance to the passage of electricity between the two electrodes produces enough heat to fuse the metal in between. This creates a spot weld. When a circular electrode is used, a continuous seam weld is formed.

Brazing and soldering are alternative methods of joining metals. Brazing involves melting brass to form a joint; soldering does the same thing, only with melted solder.

▲ The former transatlantic liner *Queen Mary*, launched in 1934, is now a floating hotel and conference centre at Long Beach, Los Angeles, USA. Like most ships of its day, the 310.8-m long vessel is built of riveted steel plates.

▶ The huge Magnus oil-production rig just after construction. It was built by welding together sections of steel pipe, which measured 13 km long overall and weighed 14,000 tonnes. It is now sited in the Magnus oilfield off the Shetland Isles of Britain.

Machining

Most metal objects shaped by casting, forging or other methods need some kind of finishing treatment before they are ready for use. For example, they may need to have holes drilled or metal removed to bring them to the right size and shape. The machines that carry out such metal-finishing processes are known as machine tools. They play an important part in the modern assembly-line method of manufacturing because they can work to very accurate limits and produce near-identical parts.

Since machine tools are used to cut metal, they have powerful motors to drive the cutting tools. These tools are made from very hard tool steels, which retain their sharpness during machining. Special high-speed steels contain-ing tungsten and chromium remain sharp even when they run red-hot. To help reduce temperatures during machining, the tool and workpiece are cooled by a light "cutting oil". This also helps lubricate the cutting operation.

The lathe

One of the most common machine tools in workshops is the lathe, on which a process called turning is carried out. On a lathe, a workpiece is rotated and various cutting tools are then moved in to cut it. The workpiece is rotated between a headstock at one end and a tailstock at the other. It is clamped in a chuck in the headstock, which also houses the motor that drives the chuck. A gearbox allows the

▲ Machinists examine a wing panel for a European Airbus, which is being cut to shape on a huge milling machine. Twin cutting heads are working on two panels. The wing panels on the Airbus are machined from solid metal. This method of construction is much stronger than conventional riveting.

▲ A lathe operator checks the diameter of a large turbine rotor with a gauge. The rotor was originally shaped on a hydraulic forging press. It is now being "turned" on a lathe, where metal will be removed until it is the right size. The chuck of the lathe is at the top; the tailstock is at the bottom.

workpiece to be driven at a number of speeds from, say, 20 to 2,000 revolutions per minute.

The cutting tools are mounted on a cross-slide, which in turn is mounted on a saddle. The saddle moves lengthwise along the lathe, while the cross-slide moves at right-angles to it, so as to carry the tools towards or away from the rotating workpiece. The cross-slide and the saddle both run on precision screw threads so that they can be positioned with great accuracy.

Drilling and milling

Another common machine tool is the drill press, which is used to drill holes. The workpiece is held stationary, while a rotating drill bit is lowered into it. The drill bit has cutting edges just at the tip and spiral grooves, or flutes, along the side. This allows the cut metal, known as swarf, to escape. Turret drills have a drill head that carries a number of drill bits of different sizes.

Milling is a machining operation carried out with a rotating toothed cutting wheel. Metal is removed as the workpiece moves past the wheel, which may rotate at speeds approaching 10,000 revolutions per minute.

Other machine tools carry out other metal-finishing operations. For example, a shaping machine uses a chisel-like tool to cut flat surfaces; a grinding machine uses a rotating abrasive wheel or a moving abrasive belt to remove metal.

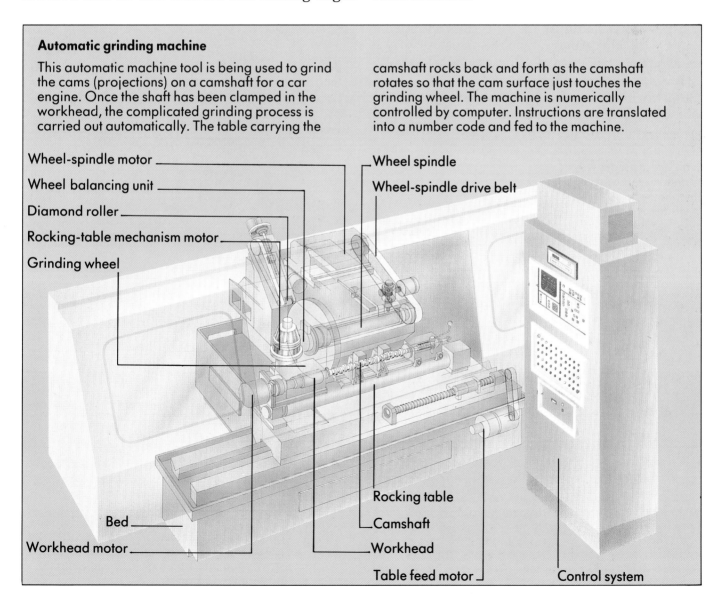

Automatic grinding machine

This automatic machine tool is being used to grind the cams (projections) on a camshaft for a car engine. Once the shaft has been clamped in the workhead, the complicated grinding process is carried out automatically. The table carrying the camshaft rocks back and forth as the camshaft rotates so that the cam surface just touches the grinding wheel. The machine is numerically controlled by computer. Instructions are translated into a number code and fed to the machine.

Wheel-spindle motor
Wheel balancing unit
Diamond roller
Rocking-table mechanism motor
Grinding wheel
Wheel spindle
Wheel-spindle drive belt
Bed
Workhead motor
Rocking table
Camshaft
Workhead
Table feed motor
Control system

Iron into steel

Pig iron is produced when iron ore is smelted with coke and limestone in a blast furnace. It is not pure iron, but contains a lot of impurities, particularly carbon (about 4 per cent). This makes the iron brittle. Only when most of the carbon is removed does the metal become really useful. It then becomes steel.

Steel is the name we give to the alloy, or mixture, of iron with traces of carbon. The presence of just a few parts per thousand of carbon makes iron much stronger and harder than it is when pure.

In steelmaking, the excess carbon is literally burned out of the pig iron in a furnace. Most steel is now made by the basic-oxygen method. The carbon burns off when a jet of oxygen is blasted into the molten pig iron. The highest quality steel is made by melting selected steel scrap in an electric-arc furnace. Heat is produced in this furnace by an electric arc: a kind of continuous electric spark.

Steelworks are vast. They not only produce the metal, but also carry out many shaping processes, such as rolling, forging and casting.

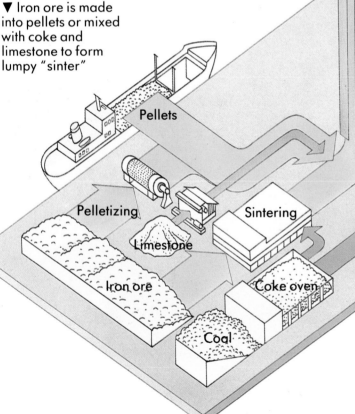

▼ Iron ore is made into pellets or mixed with coke and limestone to form lumpy "sinter"

Pellets

Pelletizing

Limestone

Iron ore

Sintering

Coke oven

Coal

Coke

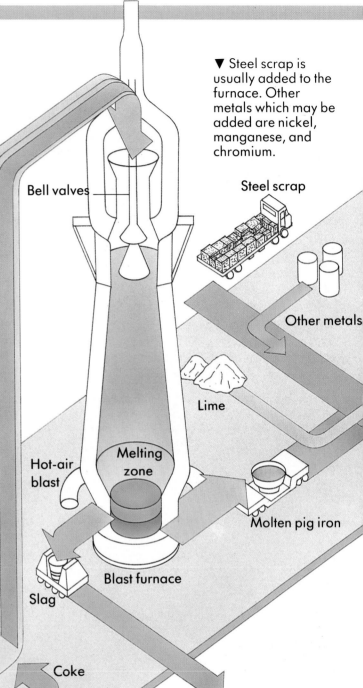

▼ Steel scrap is usually added to the furnace. Other metals which may be added are nickel, manganese, and chromium.

Bell valves

Steel scrap

Other metals

Lime

Hot-air blast

Melting zone

Molten pig iron

Slag

Blast furnace

▲ The blast furnace is a steel tower, standing about 60 m high and measuring 10 m in diameter. Iron ore, coke and limestone are charged into the top of the furnace through a "double-bell" valve system. This prevents the loss of the furnace gases, which include carbon monoxide. The gases are burned as fuel in stoves. These stoves heat the air which is blasted into the base of the furnace.

94

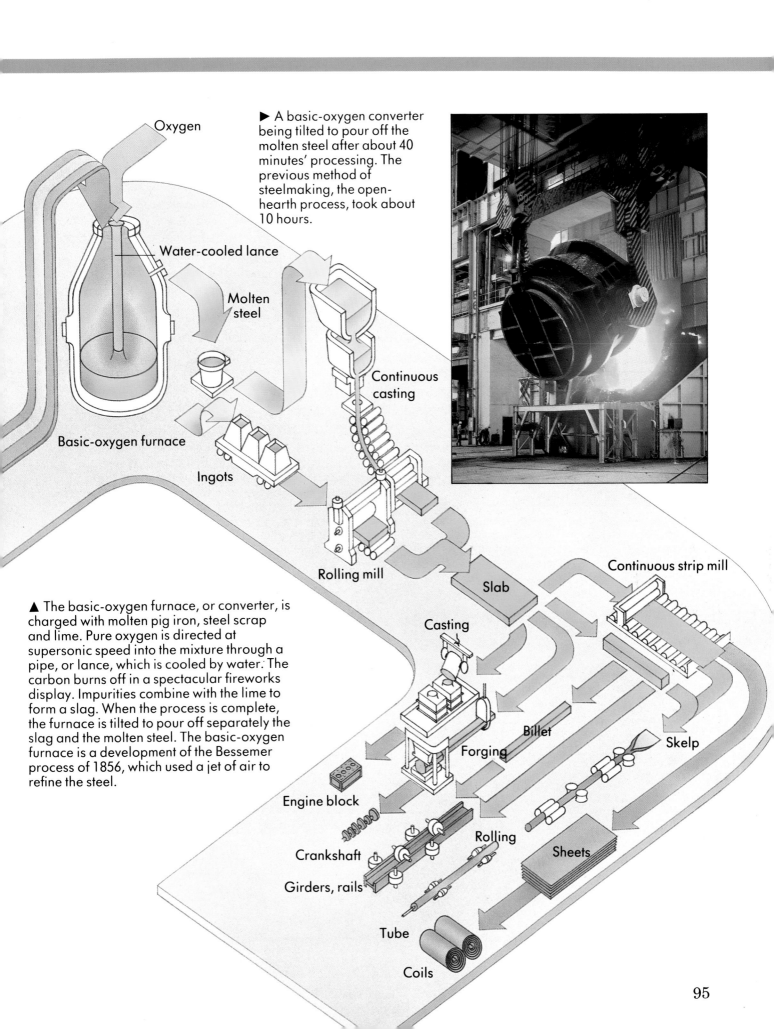

Oxygen

Water-cooled lance

Molten steel

Basic-oxygen furnace

Ingots

Rolling mill

Slab

Continuous casting

Casting

Forging

Billet

Skelp

Continuous strip mill

Engine block

Crankshaft

Girders, rails

Rolling

Sheets

Tube

Coils

▶ A basic-oxygen converter being tilted to pour off the molten steel after about 40 minutes' processing. The previous method of steelmaking, the open-hearth process, took about 10 hours.

▲ The basic-oxygen furnace, or converter, is charged with molten pig iron, steel scrap and lime. Pure oxygen is directed at supersonic speed into the mixture through a pipe, or lance, which is cooled by water. The carbon burns off in a spectacular fireworks display. Impurities combine with the lime to form a slag. When the process is complete, the furnace is tilted to pour off separately the slag and the molten steel. The basic-oxygen furnace is a development of the Bessemer process of 1856, which used a jet of air to refine the steel.

95

Producing other metals

Lead, zinc and tin are smelted in blast furnaces in a similar way to iron. Some metals are produced in different kinds of furnaces. Others are extracted by means of electrolysis: this is known as electrometallurgy. Electrolysis is also widely used for purifying metals that may have been extracted by other methods. The extraction of copper from different ores provides a good illustration of these alternative methods.

Copper smelting

Copper often occurs in sulphide ores, in which it is combined with other metals, particularly iron and nickel. The presence of these other metals complicates the extraction process. The follow-ing smelting processes are used for the ore chalcopyrite, or copper pyrites, which is a mixed sulphide of copper and iron.

The ore is first concentrated by flotation and then smelted in a reverberatory furnace. Flames shoot over the concentrate and turn it into a bubbling, boiling mass. Some of the iron ore and the earthy impurities together form a slag, which is run off. What remains is matte, a mixture of copper and iron sulphides.

The matte is transferred to another furnace, called a converter, and air is blown through it. Sand (silica) is added, which absorbs the iron and other impurities. This results in blister copper metal, which is about 98 per cent pure.

Smelting aluminium

Aluminium is produced by the electrolysis of molten aluminium oxide. The principle is simple, but the practice is complicated. First, the aluminium oxide must be extracted from bauxite, the ore of aluminium.

This is done by the Bayer process, in which the bauxite is digested with caustic soda (sodium hydroxide). The aluminium oxide dissolves to form sodium aluminate. Crystals of aluminium hydroxide form when the aluminate is cooled. These are filtered off and then heated. This process (calcination) produces alumina (aluminium oxide).

Alumina by itself does not melt until it reaches about 2,000°C. Mixed with a mineral called cryolite, it will melt at only about 1,000°C. In aluminium smelting, a mixture of alumina and cryolite is charged into the furnace. In the furnace, carbon rods (the anodes) are lowered into the molten mixture. Electricity passes between them and the carbon furnace lining (the cathode). The electricity splits up the aluminium oxide into aluminium metal, which collects as a molten layer on the floor of the furnace. Oxygen combines with the carbon anodes to form carbon monoxide, which is led off.

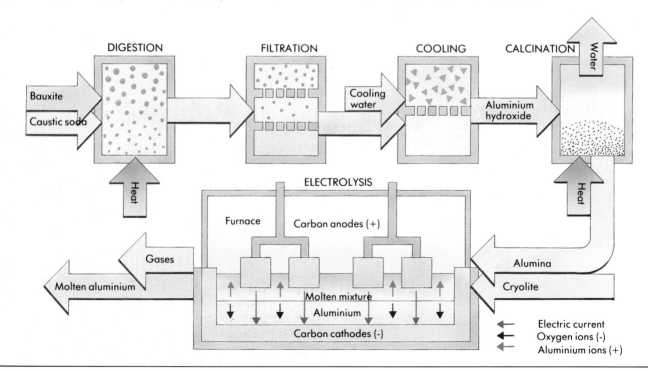

Electrolysis

The purification of the copper is completed by electrolysis: passing electricity through a solution. The impure copper is made into blocks, which become the anodes (positive electrodes) in the process. Sheets of pure copper become the cathodes (negative electrodes). They are placed in a bath of copper sulphate solution, and an electric current is passed through them.

Under the influence of the electricity, copper from the impure anode goes into solution as ions (charged atoms). Pure copper comes out of solution at the cathode, where copper ions change back into atoms. The result is that the anodes dissolve away, while the cathodes grow. Impurities settle out as a slime.

Leaching

The electrolysis of copper sulphate also features in the extraction of copper from oxide ores, such as cuprite. The ores are treated with sulphuric acid, which dissolves the copper as copper sulphate. This kind of process is called leaching. It is an example of hydrometallurgy, the extraction of metals by means of chemical solutions.

Leaching is also an important method of removing uranium, gold and silver from low-grade ores, which contain only the minutest amounts of metal. Uranium is also extracted by treatment with sulphuric acid. Gold and silver are removed from their ores using a weak solution of sodium cyanide.

▲ Molten zinc being tapped from a furnace. Zinc is smelted in a blast furnace, which it leaves as a vapour because of its quite low melting point. Molten metal forms when the vapour is cooled.

▼ The copper in some ores is removed by treatment with sulphuric acid (left). The copper sulphate that forms is then reduced to copper metal by electrolysis. Copper is deposited on the cathode plates of the electrolytic bath (below), which holds the copper sulphate solution. It is deposited as very pure copper.

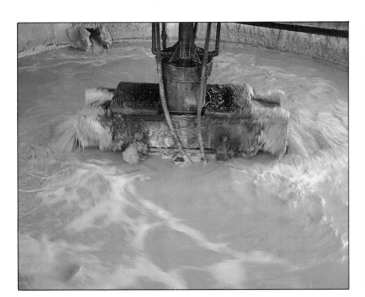

The chemical industry

Spot facts

• *The two most widely used industrial chemicals are sulphuric acid and ammonia.*

• *A large chemical plant producing caustic soda by the electrolysis of brine uses as much electricity as a town of 300,000 people.*

• *Chemists have synthesized as many as 60,000 different kinds of plastics.*

• *Polyethylene plastic is made up of molecules that have up to 20,000 carbon atoms linked together in a long chain.*

• *A pair of nylon stockings is made up of more than 6 km of yarn knitted into three million loops.*

► A chemical engineer adjusts a valve regulating the flow of materials through a pilot plant. This is a small-scale chemical plant used for testing a new process. If it performs satisfactorily, a full-scale plant might be built.

From its beginnings in the 1700s, the chemical industry has grown into one of the largest industries there is. Its products, chemicals, are used in practically every other industry, from electronics to steelmaking. They are used in agriculture, for fertilizers and pesticides, and in the home, in paints, washing powders, hairsprays and medicines. And most of us use, wear, and even eat the products made from, or processed with, chemicals. These range from margarine and paper to drip-dry shirts and non-stick cookware.

The chemical industry uses processes devised in the chemistry laboratory. It transforms raw materials into finished chemical products, or else into intermediate chemicals that other manufacturers turn into products.

Chemical engineering

Chemical engineering is the branch of engineering which designs, builds and operates chemical plants, the factories in which chemicals are produced or processed.

Chemical engineers take a process from the chemistry laboratory and develop it on a large enough scale for industrial production. This is usually much more difficult than it sounds. For example, in the laboratory, heating a few millilitres of acid in a glass flask over a Bunsen burner presents few problems. But heating up thousands of litres of acid in an industrial plant is quite a different matter.

Each chemical manufacturing process uses raw materials and converts them into finished products. In this way, each process is different. However, every process involves a certain number of standard chemical and physical operations. Chemical engineers design suitable equipment to carry out these operations economically. Where possible, they use standard, rather than specially designed equipment, to keep down the cost.

Standard chemical operations are known as unit processes. They include oxidation, chlorination and hydrogenation, in which, respectively, oxygen, chlorine and hydrogen combine with other substances. Another important unit process is polymerization, in which small molecules are built up into larger ones.

Standard physical operations are known as unit operations. Common ones include mixing, filtering, distillation, evaporation and drying.

▶ A chemist works in a laboratory on a method of vacuum-processing materials. There is no guarantee that a large-scale plant would be able to repeat the process successfully or economically.

▼ The diagram shows a likely timescale for the nine main stages from the birth of an idea to the production of a saleable product. The whole process often takes more than five years from start to finish. It could take another five years for the plant to pay for itself.

| Project formulated | Process discovered | Process defined in laboratory | Small test plant built | Process defined in test plant | Sanction for full scale plant | Full scale plant designed and constructed | Plant commission and start-up | Plant on stream |

Time (months)

| 6 | 12 | 18 | 24 | 30 | 36 | 42 | 48 | 54 | 60 |

Heavy chemicals

The chemical industry produces vast tonnages of a wide range of chemicals. But a relatively small number of chemicals account for the bulk of production. These are generally called heavy chemicals, because they are produced in such large amounts. In contrast, some chemicals are produced only in small amounts. They are usually termed fine chemicals. They are also often the product of more complex chemical processing. Dyes and pharmaceuticals, or drugs, are examples of fine chemicals.

Most of the leading heavy chemicals are inorganic. They are made from salts, minerals or gases in the air. Among the most important are sulphuric acid, ammonia, sodium hydroxide and sodium carbonate. Sulphuric acid is vital to so many modern manufacturing processes that it is often called the "lifeblood of industry". But the other three chemicals mentioned are also vital to modern industry.

Early chemical industry

The modern chemical industry began in the 1790s. That is when Nicolas Leblanc, a French surgeon turned chemist, found a way of making sodium carbonate on an industrial scale. The chemical was much in demand for making soap and glass. The first stage of the Leblanc process was to treat sodium chloride (common salt) with sulphuric acid. The demand for sulphuric acid for the Leblanc process led in turn to an improved process for making the acid, called the lead-chamber process. This was later superseded by the present method, which is called the contact process.

Salt water is the starting point for the modern method of making sodium carbonate. This is called the ammonia-soda process because it involves a series of reactions with ammonia. Salt water is also the raw material for making caustic soda, or sodium hydroxide. But this time no lengthy series of chemical reactions is involved. Caustic soda is produced simply by passing an electric current through the salt water. This method, electrolysis, is a useful way of producing many metals and chemicals.

▶ This plant makes ammonia by combining nitrogen and hydrogen in the presence of an iron-oxide catalyst. The process, called the Haber synthesis, takes place at about 400°C and at a pressure of up to 1,000 atmospheres.

Making sulphuric acid

Sulphur is the usual starting point in the manufacture of sulphuric acid. It is heated with air in a furnace and oxidized to sulphur dioxide gas. After being cooled in a heat exchanger, the gas is fed to a converter. There, with the help of a catalyst, it is further oxidized to sulphur trioxide gas. This gas is absorbed by a spray of dilute sulphuric acid. Concentrated acid results.

Heat exchange — Furnace

Sulphur

Dry air

Steam

Water

SO_2

SO_2

Converter
Absorber

SO_3

Waste gas

Sulphuric acid

Vent to atmosphere

Concentrated Sulphuric acid

► The main uses of four of the world's leading industrial chemicals. A major use of sulphuric acid and ammonia is to make fertilizers. The acid is used to make superphosphate; ammonia is used to make ammonium nitrate and urea. Caustic soda, or sodium hydroxide, is used for making soap, paper and artificial silks. One of sodium carbonate's most useful applications is in the manufacture of glass.

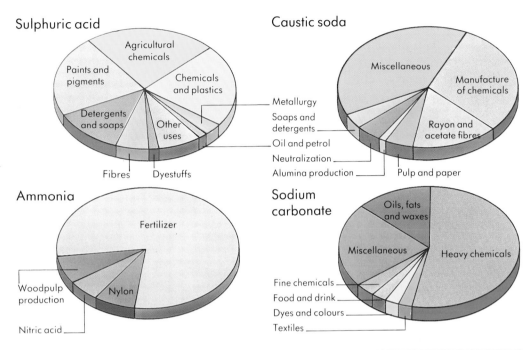

Sulphuric acid

Agricultural chemicals
Paints and pigments
Chemicals and plastics
Metallurgy
Detergents and soaps
Other uses
Fibres
Dyestuffs

Caustic soda

Miscellaneous
Manufacture of chemicals
Soaps and detergents
Rayon and acetate fibres
Oil and petrol
Neutralization
Alumina production
Pulp and paper

Ammonia

Fertilizer
Woodpulp production
Nylon
Nitric acid

Sodium carbonate

Oils, fats and waxes
Miscellaneous
Heavy chemicals
Fine chemicals
Food and drink
Dyes and colours
Textiles

Plastics

Plastics are now close to metals in being the most important industrial products of our age. We can define a plastic as a substance which has a long-chain molecule, and which can be moulded into shape when heated.

It is the long-chain molecules of plastics that make them so special. Most ordinary substances have short molecules, with just a few atoms linked together. Plastics, however, have long molecules containing thousands of atoms, almost always with a "backbone" of linked carbon atoms. Carbon is the only element that can link together in this way.

The raw materials for most plastics are hydrocarbons obtained from oil refining. The most useful of these is the gas ethylene (ethene). This has a short molecule with a backbone of just two carbon atoms. However, at high temperature and pressure, thousands of the short ethylene molecules will link together to form a long-chain molecule. We call ethylene a monomer ("one part"); the long-chain molecule, a polymer ("many parts"); and the process, polymerization. We know this particular polymer as polyethylene, or the material polythene.

Thousands of plastics can be produced by the polymerization of suitable hydrocarbons or their derivatives. Among other well-known plastics are PVC (polyvinyl chloride), nylon, polypropylene and polystyrene. All these plastics will soften when reheated. They are known as thermosoftening plastics, or thermoplastics. The other main group of plastics set rigid when they are heat-moulded into shape and will not soften when reheated. They are called thermosetting plastics, or thermosets. They include the original synthetic plastic, bakelite (phenol-formaldehyde), and its relatives urea- and melamine-formaldehyde.

▲ Like other modern cars, this MGB Roadster uses many different kinds of plastics. The tyres are made from synthetic rubber, as are the shock-absorbing bumpers. The paint was made using plastic resins, while the upholstery and carpets are woven from synthetic fibres. The hood is made from PVC, textured to imitate leather.

▶ A reservoir on Tenerife, in the Canary Islands, which has been lined with PVC.

Synthetic rubber

Many products used today are made from synthetic rubbers. They are a kind of "elastic plastic", a material called an elastomer. The search for a substitute for rubber led German chemists to produce the first successful synthetic rubber in 1927. Called buna rubber, it was made from butadiene, a chemical closely related to isoprene, the monomer in the sap of the rubber tree.

The most common synthetic rubber today is a copolymer (mixed polymer) of butadiene and styrene. Neoprene is a synthetic rubber made from acetylene (ethyne). It was one of the first to be discovered, by Wallace H. Carothers in 1928, and is still widely used because of its excellent resistance to high temperatures, oils and chemicals.

103

Shaping plastics

By far the commonest methods of shaping plastics involve moulding. Thermoplastics such as polyethylene and PVC are easy to mould into shape, and various methods are possible. Bowls, for example, are made by injection moulding. This involves heating the plastic until it is molten and then injecting it into a shaped, water-cooled mould.

Bottles and hollow toys can be made by blow moulding. A blob of molten plastic is delivered into a hollow mould, and then air is blown into it through a pipe. The plastic is forced against the mould and takes its shape.

Thermosetting plastics, such as bakelite, have to be shaped by a different technique, called compression moulding. They cannot be shaped like thermoplastics because they melt and set more or less at the same time. During their manufacture, the polymerization process is halted before the molecules begin to crosslink and set hard. This produces a so-called moulding resin. Objects such as cups are shaped when this resin is simultaneously heated and compressed in a mould.

Plastics can also be shaped by extrusion and laminating. Pipes, for example, are made by extrusion. A screw-like device forces molten plastic through the hole in a die. Plastic sheet is made by extruding molten plastic through a ring-shaped slit. Heatproof surfaces are made by laminating: sandwiching together layers of material soaked in thermosetting plastic resin.

Vacuum forming

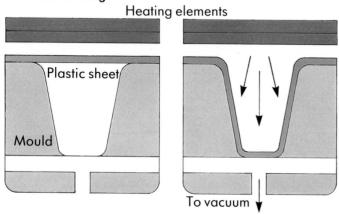

A sheet of plastic is placed on top of the mould and heated until it is soft. The mould is then connected to a vacuum line, and the air is sucked out of it. Outside air pressure forces the plastic into the mould.

Blow moulding

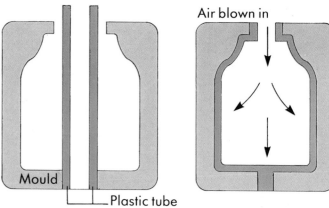

A length of hot plastic tubing is placed in the open mould. This then closes and seals the bottom. Air is blown into the tube from above, forcing the plastic against the walls of the mould.

Injection moulding

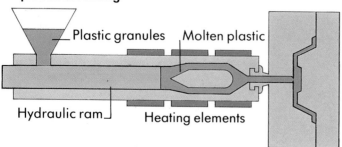

Plastic granules are fed into the injection-moulding machine and heated until they melt. A hydraulic ram then forces the molten plastic into the water-cooled mould, where it cools and sets.

Extrusion

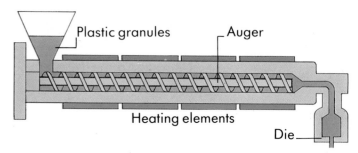

In the extrusion machine, plastic granules are heated until they melt. A screw-like device called an auger rotates and forces the molten plastic through a shaped hole called a die.

104

Man-made fibres

Silk, the finest natural fibre, is produced, or "spun", by the silkworm. In 1884 a French chemist, Hilaire Chardonnet, succeeded in imitating the silkworm and produced long fibres of what he called artificial silk. The material he used was cellulose nitrate. To make the fibres, he dissolved it in a solvent. He then forced the solution through the fine holes of a device similar to the spinning gland of the silkworm. Fibres formed when the solvent evaporated from the fine streams of solution.

In 1892 a better method of making artificial silk was developed, called the viscose process. It produced fibres of pure cellulose. This process is still very important today, producing fibres called viscose, or viscose rayon. The method involves treating the cellulose first with caustic soda and then with carbon disulphide. Fibres form when cellulose solution is pumped through a spinneret into an acid bath. In the bath the cellulose is regenerated. Acetate and triacetate are similar fibres made from cellulose acetates.

Many fibres used today, however, are wholly synthetic. They are kinds of plastics that can be drawn out into continuous threads, or filaments. Synthetic fibres are very strong, do not rot or absorb water, and are not attacked by insects. Among the best-known are nylon, polyester and acrylic fibres. Nylon and polyester fibres are produced by melt spinning: forcing molten plastic through a spinneret. The acrylics are produced from a solution of the plastic.

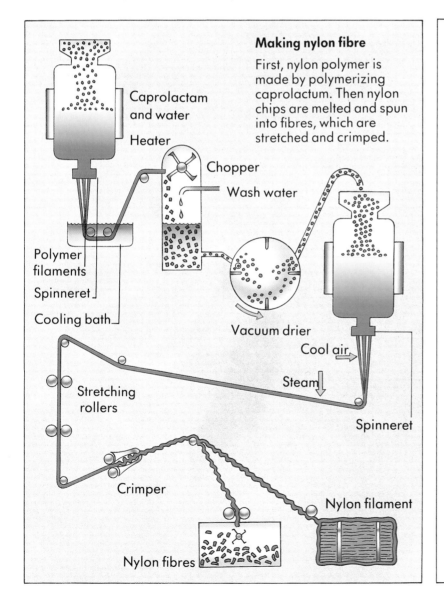

Making nylon fibre

First, nylon polymer is made by polymerizing caprolactum. Then nylon chips are melted and spun into fibres, which are stretched and crimped.

Caprolactam and water

Heater

Chopper

Wash water

Polymer filaments

Spinneret

Cooling bath

Vacuum drier

Cool air

Steam

Spinneret

Stretching rollers

Crimper

Nylon filament

Nylon fibres

The nylon man

In the early 1930s an American chemist, Wallace H. Carothers, headed a research team trying to find a substitute for silk. Carothers eventually found two coal-tar chemicals, adipic acid and hexamethylene-diamine, that would copolymerize to form molecules like those of silk. He produced the first practical fibre from the polymer in 1935. It was the first synthetic fibre, which was fine and lustrous, elastic and strong. It came to be called nylon.

105

Food and drugs

- A strain of "super rat" is now breeding which can resist pesticides such as warfarin.

- The synthetic sweetener saccharin (the cyclic imide of ortho-sulphobenzoic acid) is 550 times sweeter than ordinary sugar.

- Aspirin (acetylsalicylic acid) is the world's commonest drug. People in the United States alone take 4 tonnes of aspirin tablets every day to treat colds and headaches.

- Digitalis, a heart stimulant, is one of the oldest drugs still in use. It is prepared from the dried leaves of the purple foxglove.

Chemical processing plays a major part in the daily lives of most people in developed countries. Farmers apply chemical fertilizers to the soil to make their crops grow better and produce greater yields. They spray the crops with chemicals to kill insects and protect them from disease. Our food is often treated with chemicals so that it looks and tastes more appetizing, and can be kept for longer periods without deteriorating. Methods of food preservation enable us to enjoy a wider range of foods. Without the benefit of man-made drugs, we would succumb to all manner of diseases.

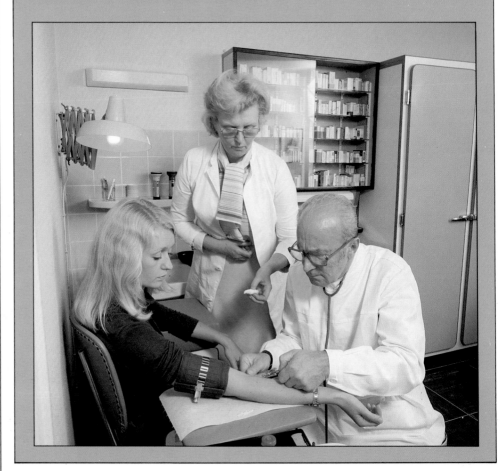

▶ A doctor injects a drug into a patient with a hypodermic syringe. The body of the syringe carries a graduated scale, allowing an accurate dose of the drug to be administered. A subcutaneous injection goes under the skin. An intravenous injection, such as this, goes into a vein.

Agricultural chemicals

When crops grow, they extract nutrients from the soil. To ensure that the soil remains fertile, these nutrients must be replaced. This is done by applying fertilizers. In the early days of farming, animal manure was enough. Nowadays chemical fertilizers, such as superphosphate and ammonium compounds, are used.

Growing crops can be attacked by many insect pests and also many fungal diseases. Again, the chemist comes to the aid of the farmer by creating powerful insecticides and fungicides. Competition with weeds is also eliminated by means of herbicides (weed-killers). Many of the most potent pesticides are chlorinated hydrocarbons, such as dieldrin and DDT. These chemicals are deadly to animal life, and once in the food chain, their effects accumulate. Other effective pesticides, including organic phosphorus compounds, are less toxic to animal life and are not so persistent.

◀ Chemical herbicide has been sprayed around this young oil palm to kill weed growth. This allows the palm room to establish itself.

▼ An Asian farmer sprays insecticide on a cereal crop to prevent insect pests breeding and multiplying.

107

Food technology

Most of the foods we eat have been processed in some way. Even fresh foods such as fruits may have been treated with chemicals to assist ripening. Food processing began thousands of years ago, when early peoples began to make bread from the grain crops they gathered. Bread is still one of the basic foodstuffs of the world, and has been called the "staff of life".

The principles of making bread have hardly changed over the years, although it is now often mass-produced in factories. Bread is made by baking a prepared dough in a hot oven. The dough is a mixture of flour, salt and water, to which yeast has been added to make it ferment. The fermenting yeast produces carbon dioxide gas, which makes the dough rise. This gives

▼ Fermentation tanks at a distillery. In these tanks sugars extracted from grain are fermented with yeast. The yeast changes the sugars into alcohol, with carbon dioxide bubbling off as a waste product.

bread its typical light texture when it is baked.

From early times also fermentation has been used for another purpose, making beer. Beer is produced by fermenting watery mixtures of grains. The yeast turns sugar extracted from the grain into alcohol, while carbon dioxide is given off as a waste product. This reaction was probably the first chemical process utilized by man.

Like bread, milk has been part of our staple diet since the beginning of civilization. And it has also been processed into other foodstuffs for nearly as long. One is butter. This is made by churning the cream that settles on top of the milk. Churning – rotating the cream in a drum – causes the little fat globules in the cream to join together into a solid mass, butter. Cheese is another food derived from milk. It is made by adding rennet to milk, which makes it set into a solid curd. This then matures into cheese.

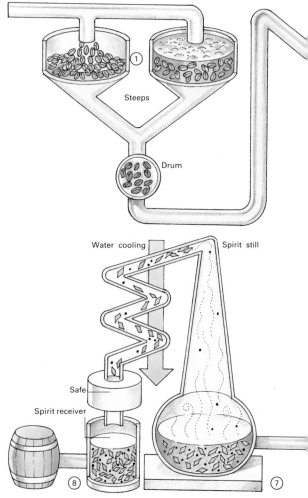

We can think of cheesemaking as a way of preserving milk, which otherwise "goes off", or turns sour, in a day or so. It would go off even sooner were it not for a treatment it receives in the dairy before bottling or packaging. This treatment is pasteurization, which involves the milk being heated briefly, then quickly cooled.

Several other methods are used to preserve food. They all aim to halt or slow down the processes that cause food to spoil. Spoilage may be brought about by microorganisms, such as bacteria, or by chemical changes. Traditional methods of preservation include smoking, pickling and drying. The commonest methods of preservation are canning and freezing.

▶ Filling cans with fish in a canning factory. The cans will next be sealed, and then sterilized by heating in batches.

Making malt whisky

The making of malt Scotch whisky is a lengthy process. The first step is malting, which involves soaking barley until it germinates, or starts to shoot (1). The malted barley is then dried in a kiln (2). After being weighed, the dried barley is ground up (3) and mixed with hot water to form mash (4). The sugar (maltose) in the grain passes into solution to form a liquid called wort. After filtering and cooling, the wort is fermented with yeast (5) for about two days. The liquid is then distilled (6), producing a weak alcohol and water solution. This solution is distilled again (7), producing a "spirit" with a high alcoholic content (8).

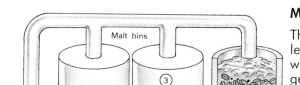

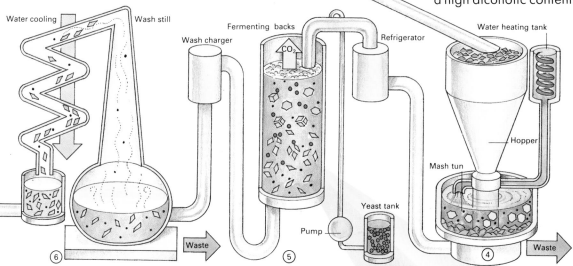

⬭ Barley
⬭ Germinated barley
⬭ Malted barley
⋮ Peat smoke
◇ Malt grist
⬡ Sugar (Maltose)
∴ Yeast
🌱 Alcohol (ethanol)

Synthetic foods

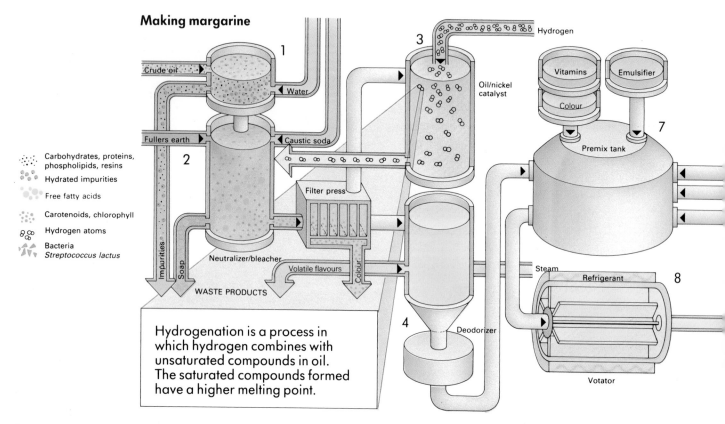

Making margarine

Crude oil
1
Water
Fullers earth
2
Caustic soda
3
Oil/nickel catalyst
Hydrogen
Vitamins
Emulsifier
Colour
7
Premix tank
Filter press
Impurities
Soap
Neutralizer/bleacher
Volatile flavours
Colour
WASTE PRODUCTS
4
Deodorizer
Steam
Refrigerant
8
Votator

Carbohydrates, proteins, phospholipids, resins
Hydrated impurities
Free fatty acids
Carotenoids, chlorophyll
Hydrogen atoms
Bacteria *Streptococcus lactus*

Hydrogenation is a process in which hydrogen combines with unsaturated compounds in oil. The saturated compounds formed have a higher melting point.

In 1869 a French chemist named Hippolyte Mège-Mouriès patented one of the first synthetic foods. He called it margarine, a name loosely based on his surname. It won him a contest launched by Emperor Napoleon III to find a palatable substitute for butter. Mège-Mouriès made his margarine using fats from beef suet, pig's stomach and cow's udder. He mixed with them skimmed milk, or whey.

Animal fats are still used to make some margarines. But most are made using vegetable oils, including safflower, sunflower, coconut and palm oils. These oils are converted during manufacture into solid fats by treatment with hydrogen. Margarines based on vegetable oils now sell very well because it is believed that they are healthier to eat than butter and other animal fat products.

Margarine contains various additives which give it the right consistency, improve its nutritional value, help preserve it and enhance its appearance. They include emulsifiers, which prevent the fats and water in the margarine separating out. An important emulsifier is lecithin, found in egg yolks. Vitamins A and D may be added to increase the food value.

The colour in margarines comes from a dye called beta-carotene, which occurs naturally in carrots. But most of this colouring is now made synthetically from coal tar. Potassium sorbate is a common preservative found in margarines.

Emulsifiers, vitamins, colouring and preservatives are common additives found in most processed foods today. Other additives include thickeners, such as gelatine and alginates (extracted from seaweed), and anti-oxidants. These are mainly synthetic compounds that stop fats going rancid and other foods developing unpleasant flavours. Monosodium glutamate is a common additive that brings out the flavour of food. Sweetness is provided by the addition of glucose or other sugars, or by synthetic sweeteners such as aspartame and saccharin.

Synthetic proteins include meat substitutes, properly called texturized vegetable protein (TVP). The vegetable involved is the soya bean. Protein is extracted from the beans and then dissolved in alkali. The solution is then extruded through a spinneret into an acid bath. The protein comes out of solution as fibres, which are gathered into a rope and then chopped.

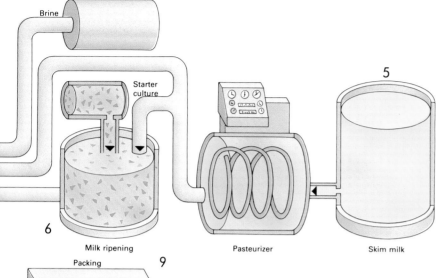

Brine

Starter culture

6

Milk ripening

Packing   9

Pasteurizer

5

Skim milk

◄ Margarine is now made mainly from vegetable oil. The impure, or crude oil is washed with water (1) and then treated with caustic soda and Fuller's earth (2). The alkali combines with unwanted substances to form a soap. The Fuller's earth removes colour. Some oils require treatment with hydrogen to make them solid (3). Steam is then bubbled through the purified oil in a deodorizer (4) to remove any remaining odour. Meanwhile, skimmed milk is being prepared in another part of the plant (5). After being pasteurized, it is "ripened" by treatment with bacteria (6). It then passes with the oil and brine into the premix tank (7). Other ingredients are also added at this stage. These may include colourings, to give a "buttery" colour; vitamins, to improve its food value; and an emulsifier. The margarine mixture solidifies in a refrigerated rotating device called a votator (8) and passes to the packing machine (9).

The food chemist

Chemists help food manufacturers ensure that their products look good, taste good, have a long shelf-life and, above all, are safe. Food chemists (inset) study, for example, how the body recognizes flavours and the mechanisms of food decay. They develop additives to improve flavour, arrest decay, and so on.

They also analyse natural flavourings and try to imitate them or improve upon them by chemical synthesis. In their analysis they often use a technique called chromatography to separate out the chemicals in a substance. An analysis of natural peppermint oil (left) shows that menthol, menthone and methyl acetate are the main ingredients. A mixture of these chemicals would produce an artificial peppermint flavour that is very similar to that of the natural product.

In many countries all additives in food products must be listed on the label. In Europe they are usually identified by an E number, which indicates that they have been approved for use in EEC (European Economic Community) countries.

Menthol

Solvent

Methyl acetate

Menthone

Iso menthone
Menthofuran

Octan-3-ol

Making drugs

Penicillin

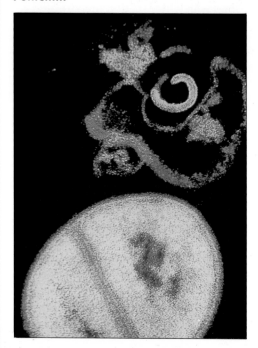

The original antibiotic, penicillin, is produced by the *Penicillium* mould. It is still widely used to combat diseases caused by such bacteria as staphylococcus, which causes boils and abcesses. The picture shows at bottom a normal staphylococcus bacterium, and at top a bacterium that has been destroyed by penicillin. Penicillin works by breaking up the bacterium's outer membrane.

◄ Traditionally, drugs were prepared using minerals and plant extracts. Some such as tincture of iodine and digitalis (from foxgloves), are still used. But most medicines are now manufactured.

▲ Opium poppies growing in Thailand. It is harvested by cutting the pods and collecting the liquid that oozes out (inset). The pain-killing properties of opium have been known for over 2,000 years.

► A technician tends a fermentation vessel in a biotechnology plant, which produces disease-fighting antibodies from cultured cells. These antibodies are used to help diagnose diseases.

Today doctors have at their disposal a vast array of medicines, or drugs, with which they can successfully treat most of the diseases that affect us. Drugs are also known as pharmaceuticals. They may be derived from plants, animals and minerals, or made from chemicals.

Opium, obtained from poppy seeds, is probably the oldest known effective drug. Opium and drugs made from it – codeine, morphine and heroin – are powerful analgesics, or pain-relievers. But they are highly addictive. This means that people who take them regularly develop a craving for them, and find it very hard to stop taking them.

Quinine is a well-known plant drug, used to treat malaria. It was obtained originally from the bark of the cinchona tree of South America. But most is now manufactured synthetically.

The manufacture of synthetic drugs dates from the late 1890s, when the Bayer chemical company in Germany began manufacturing aspirin on a large scale from coal-tar chemicals. Among other powerful synthetic drugs, the sulphonamides, or sulpha drugs, are outstanding. They fight many bacterial infections.

Among drugs obtained from animals, the hormone insulin is best known. It is extracted from the pancreas of cattle and pigs, and is used to treat diabetes.

Antibiotics are perhaps the most powerful weapons against disease. Alexander Fleming discovered the original antibiotic, penicillin, in 1928. It went into widespread production in the early 1940s. The antibiotics can now treat diseases such as pneumonia and typhoid, which in the past were generally fatal.

Everyday industries

Spot facts

● Books printed in the 1400s on parchment made from animal skin are in better condition than books printed early this century on woodpulp paper, which tends to rot.

● The latest rotor-spinning machines can produce 150 m of spun yarn per minute.

● The English chemist William H. Perkin made the first synthetic dye in 1856, while trying to synthesize the anti-malaria drug quinine.

● Today's silicon chips may contain up to half a million transistors and other electronic components. Yet they are only about the same size as a single transistor of the 1950s.

The invention of paper in China in about AD 105 can be considered a key invention in the development of civilization. It provided, in the course of time, our first means of mass communication and mass education through books, newspapers and magazines.

Spinning and weaving are two of the most ancient crafts, dating back at least 10,000 years. Until about 200 years ago, they were practised at home on simple machines, such as the spinning wheel and hand loom. In many countries, they still are. Then in the 1700s more productive spinning and weaving machines were invented. These took textile-making out of the home, and it became the first factory industry.

► Machines in a modern factory work automatically under the control of a computer, or "electronic brain". Individual machines may even have their own "brains", in the guise of a silicon chip.

Pulp and paper

Until about 150 years ago, most paper was made from linen and rags, materials that are still used to make the best-quality writing paper. But increasing demand for paper, for books and newspapers, led to wood in the form of woodpulp becoming the main raw material.

Woodpulp is made mainly from softwoods such as pine and spruces. It can be made in two ways. Mechanical pulp is made by shredding logs in a huge grinding machine. This results in a coarse pulp suitable only for making newsprint, the paper on which newspapers are printed. Chemical pulp is used to produce better quality paper. It is made by "digesting" wood chips in a solution of chemicals, usually sodium sulphate. The chemical treatment frees the wood fibres from their binder (lignin).

Woodpulp is usually transported to the papermill as dry bales. And so the first stage in papermaking is to mix the pulp with water to convert it back into a liquid state. The liquid pulp then feeds into a machine in which sets of revolving knives beat and fray the wood fibres. This enables them to bind together better later.

Next, the beaten pulp goes into a mixer tank, where it is blended with materials that will determine the quality and appearance of the finished paper. They include a filler, such as china clay, to give the paper "body" and make it smoother; size or resin glue to make the paper easier to write and print on; and maybe dyes or pigments to add colour.

The prepared watery pulp then passes to the papermaking machine, the Fourdrinier machine. It flows on to a wire-mesh belt, where the water drains or is sucked away. The damp web that forms is then squeezed by heavy rollers before being fed round steam-heated cylinders to dry. After a final rolling by heavy calendar rolls, the paper is wound on to reels.

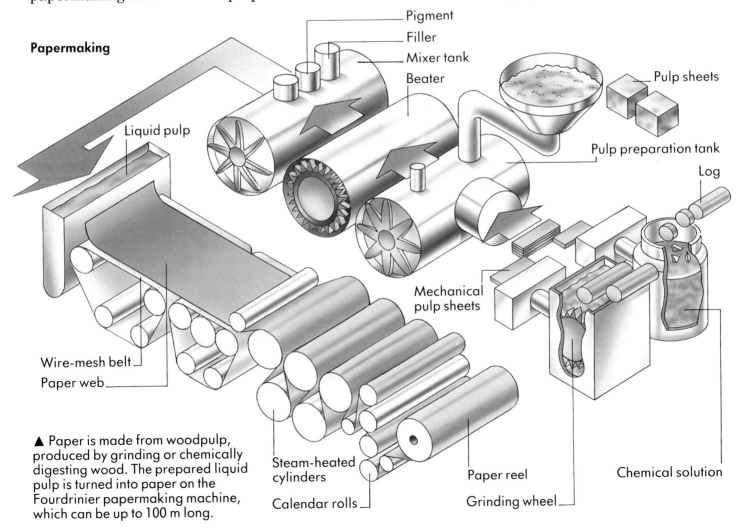

Papermaking

Pigment
Filler
Mixer tank
Beater
Pulp sheets
Pulp preparation tank
Log
Liquid pulp
Mechanical pulp sheets
Chemical solution
Wire-mesh belt
Paper web
Steam-heated cylinders
Calendar rolls
Paper reel
Grinding wheel

▲ Paper is made from woodpulp, produced by grinding or chemically digesting wood. The prepared liquid pulp is turned into paper on the Fourdrinier papermaking machine, which can be up to 100 m long.

115

Textiles

Textiles are any materials made from fibres. The most common material is cloth, made by weaving long threads, called yarns. Yarn is made by drawing out and twisting "ropes" of fibres, a process called spinning.

The traditional fibres for making cloth come from animals and plants. More and more these days, however, synthetic fibres are used instead. They are made by processing natural materials, such as cellulose, or are manufactured wholly from chemicals.

The original fibre used was wool, which comes from the fleece of sheep. It is a kind of hair that is naturally curly. The Merino breed produces the finest wool and the heaviest fleece. The fleeces of some goats, such as the Angora and Cashmere breeds, also yield excellent fibres.

Another prized animal fibre, silk, has quite a different origin. It is produced by the silkworm, the larva stage of a moth. Unlike other natural fibres, which are short, silk is produced as a continuous thread, or filament.

Cotton is by far the most important plant fibre, obtained from the boll, or fluffy seed-head of the cotton plant. Flax is a grass-like plant that has fibres in its stem. They are made into the fabric we call linen. Other natural fibres include jute and asbestos.

Spinning

With the exception of silk, the fibres from plants and animals are relatively short, usually just a few centimetres long. To make them suitable for making textiles, they must be spun into continuous yarn. Before the actual spinning process can begin, the fibres must be carefully prepared.

In the case of cotton, the bales are first opened out and the fluffy bolls are broken down into a loose fibre blanket called lap. This is then fed into a carding engine, which removes the very short fibres and also straightens out the long ones. For the best-quality yarn the fibres are straightened further by combing. They emerge from the combing machine as a web, which is then gathered into a loose rope, called sliver. Several slivers are combined and drawn out through rotating rollers to form roving. The roving goes to the spinning frames, where they are drawn out and given a twist for strength.

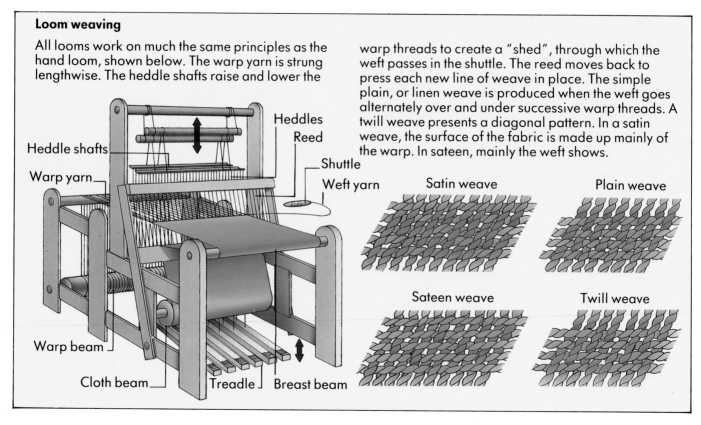

Loom weaving

All looms work on much the same principles as the hand loom, shown below. The warp yarn is strung lengthwise. The heddle shafts raise and lower the warp threads to create a "shed", through which the weft passes in the shuttle. The reed moves back to press each new line of weave in place. The simple plain, or linen weave is produced when the weft goes alternately over and under successive warp threads. A twill weave presents a diagonal pattern. In a satin weave, the surface of the fabric is made up mainly of the warp. In sateen, mainly the weft shows.

Heddles
Reed
Heddle shafts
Warp yarn
Shuttle
Weft yarn
Warp beam
Cloth beam
Treadle
Breast beam

Satin weave
Plain weave
Sateen weave
Twill weave

Weaving

Weaving takes place on a loom, on which one set of threads (the warp) is stretched lengthwise on a frame. The weaving process consists of passing thread (the weft) crosswise through a gap (the shed), created by raising and lowering sets of warp threads. Different patterns of weave are produced according to how the warp threads are separated. On traditional looms the weft is carried through the warp in a shuttle. But in the latest looms, rapier-like rods and even jets of air or water are used to carry the weft. On some of these looms, over 400 m of weft can be put down each minute.

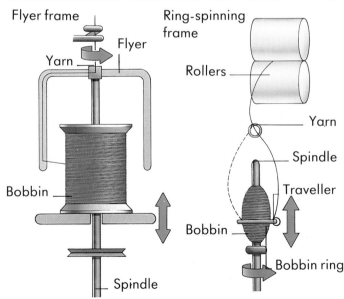

Flyer frame / Flyer / Yarn / Bobbin / Spindle — Ring-spinning frame / Rollers / Yarn / Spindle / Traveller / Bobbin / Bobbin ring

▲ (top) The final stage of spinning worsted yarn. Worsted is produced using only long wool fibres. Two common spinning machines are the flyer and ring-spinning frames (bottom). On the flyer frame, the yarn is twisted as the flyer spins round at high speed, dragging the bobbin with it. On the ring-spinning frame drawn-out yarn is wound on to the bobbin and given a twist as the traveller moves round the ring.

Dazzling dyes

Textiles have been made more attractive by dyeing for at least 5,000 years. Until the mid-1800s textile manufacturers had to rely on natural dyes, extracted mainly from plants.

In 1856, however, the English chemist William H. Perkin accidentally produced a new dye while experimenting with aniline, a liquid extracted then from coal tar. He called the new dye mauveine, which had the colour we now call mauve (see picture). It was the first synthetic dye.

Transport

The revolution in industry began in textile manufacturing, and gathered pace at the end of the 1700s. Throughout the 1800s, transport also underwent a revolution. First came the railways, on which new locomotives could haul enormous loads. These locomotives had engines powered by steam. Steam engines also took to the water, making shipping faster and more reliable.

By the end of the 1800s, road transport began to improve, with the development of the motor car. Air transport, in the form of balloons and airships, had also begun. But the real breakthrough in air transport came when the Wright brothers in the United States built the first power-driven aeroplane in 1903.

Transport by water tends to be very slow, because of friction with the water. In general, ships today are not much faster than they were a century ago. Hovercraft and hydrofoils are new kinds of vessels which have been successful in increasing the speed of transport over water. They are only used on short trips, however.

In traditional shipbuilding, hulls are usually constructed from steel plates welded together. The hull is not built up plate by plate. Large sections are built first and then put together afterwards. Glass-reinforced plastic and even concrete hulls are also made.

Railway construction has not changed much since the pioneering days of the last century. The idea of running a steel-wheeled vehicle on a steel track is a good one because of the low friction between them. Modern rails are made by rolling steel slabs with grooved rollers. The biggest change in railway practice since the last century is in the power source for the locomotives. Most are now powered by diesel engines or electricity.

The motor car, however, has changed almost

The international Airbus

The succesful European Airbus is manufactured piecemeal in plants spread throughout Europe. In Britain, British Aerospace makes the wings (left); in France, SNECMA assembles American-designed engines and Aérospatiale makes the nose and cockpit; in Germany MBB makes the fuselage and tail fin; in the Netherlands, Fokker makes the movable surfaces on the wings (flaps, airbrakes and so on); and in Spain, CASA makes the tailplane.

All the prefabricated units are transported to Aérospatiale's plant at Toulouse in south-west France, where they are assembled on a production line (right). The larger parts are delivered by the bulbous Super Guppy.

beyond recognition. Its success of the car has been a triumph for manufacturing industry. The early cars were built by hand in small numbers. Today's cars are produced in their millions. Mass-production was pioneered by car-makers. They were the first to introduce assembly lines, automation and robots.

Whereas most ships, locomotives and cars are constructed mainly in steel, most aircraft are built of aluminium alloys. These alloys are as strong as steel, but very much lighter. The aluminium sheets and other structures in the airframe, or aircraft body, are put together mainly by riveting. The method of construction, termed "fail-safe", uses staggered joints and other devices to prevent dangerous cracks running through the structure. Synthetic adhesives are also being used in airframe construction. Even the fuselage of some aircraft is now made of synthetic composites.

► Robot machines weld together steel sections to form the body shells on a car production line. Robot welders work with greater precision than humans, and are not affected by the heat and glare.

▼ The main hull and deck structure of a ship nearing completion at Ancona in Italy. Like all big ships, it is constructed of welded steel plates. When the hull is finished and painted, the ship will be ready to launch.

Electronics

Electronics is concerned with devices that control the flow of electrons in substances. It puts electrons to work in various ways. For example, it makes them create pictures on a TV screen, carry out sums in a calculator, work a computer, guide a robot and play a compact disc.

Electronics deals with the flow of electricity not so much through wires, but through substances that hardly conduct electricity at all. We call them semiconductors. The most important semiconductor by far is silicon. This element does not conduct electricity at all when it is pure, but it does – a little – when tiny amounts of impurities are added to it.

By adding different impurities to it, silicon can be given two electrical states, called n-type and p-type. By linking bits of the two types together, electronic devices like transistors, capacitors and resistors can be made. And they can be linked, or integrated, into circuits that can work TVs, calculators or computers.

There has been a revolution in electronics during the last few decades because these integrated circuits can now be made microscopically small. A tiny wafer of silicon the size of a shirt-button can carry hundreds of thousands of components and by itself run computers and other electronic equipment. We call these wafers silicon chips or microchips. A typical chip is only about 6 mm square and about one-tenth of a millimetre thick, and it weighs less than one-hundredth of a gram.

Making chips
A chip is designed so that its components and circuits can be built up in layers. There are layers of n-type and p-type silicon; a number of

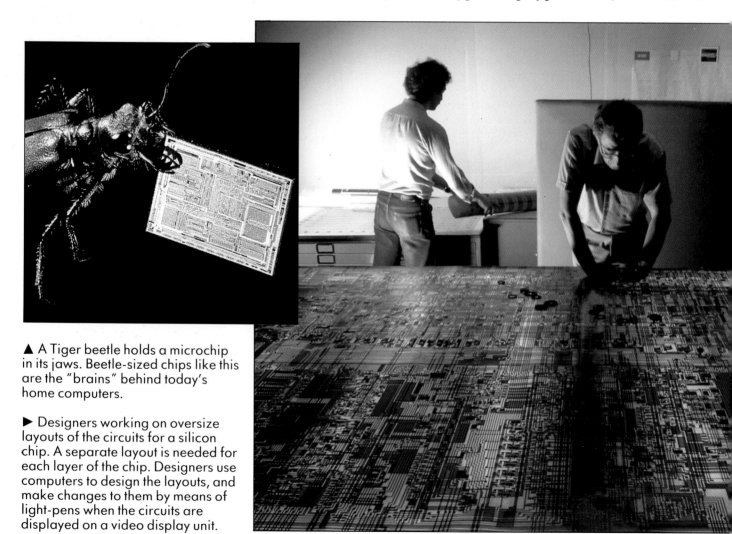

▲ A Tiger beetle holds a microchip in its jaws. Beetle-sized chips like this are the "brains" behind today's home computers.

▶ Designers working on oversize layouts of the circuits for a silicon chip. A separate layout is needed for each layer of the chip. Designers use computers to design the layouts, and make changes to them by means of light-pens when the circuits are displayed on a video display unit.

conducting and insulating layers; and a final metallic layer (usually aluminium) to provide connections. The components and pathways between them are required only in certain parts of each layer. So the other areas have to be masked off. Masks are made for each layer by photographically reducing a circuit layout 250 times its real size.

The starting point for making chips is a slice of ultrapure silicon crystal about 15 cm across. This has space for several hundred chips. The first stage of processing is to treat, or "dope", the slice with chemical vapour (often boron) to create p-type silicon. The slice is then heated in a steam oven to give it an insulating layer of silicon dioxide.

A series of photographic masking and etching processes then follow, one for each layer. They create "windows" through which the silicon can be treated. In the first masking stage, for example, areas of silicon dioxide are stripped away to allow doping of the silicon by phosphorus vapour, which creates n-type regions.

In all, more than two dozen stages of masking, doping, etching and so on, are required in making chips. Afterwards, each chip on the slice is carefully tested and inspected. As many as one in four may be rejected as a result.

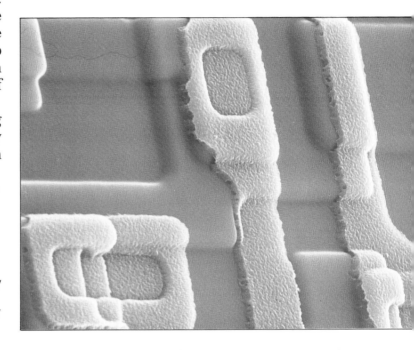

► A small section of the circuitry on a silicon chip, magnified about 4,000 times.

▼ Slices of silicon containing finished chips being inspected under the microscope. Next, they will be cut up, and the good chips will go for mounting. Many will be rejected.

▲ Computers and other electronic equipment are assembled from circuit boards like this. The various electronic components are mounted on an insulated board and connected by printed circuits, made up of films of copper. This method of assembly simplifies fault-finding.

On the road

- *The Italian artist and inventor Leonardo da Vinci designed a bicycle of sorts five centuries ago. Like many of his ideas, such as the parachute and helicopter, it was way ahead of its time.*

- *Some specially-built motorcycles called dragsters, built for fast acceleration, are able to reach speeds of more than 300 km/h from a standing start in less than seven seconds.*

- *The most successful car ever produced was the Volkswagen "Beetle", introduced in Germany in 1938. Twenty-one million vehicles were sold worldwide during the half-century it remained in production.*

- *Racing cars are fitted with upside-down "wings" (inverted aerofoils) to prevent their bodies and wheels lifting at high speeds.*

▶ Cars slow to a crawl during the evening rush hour in Sydney, Australia. A similar thing happens in most other large cities the world over, causing frustration, wasting energy and harming the environment.

The wheel is perhaps the greatest invention made by humankind. Transport on land hinges upon the wheel and modifications of it, such as the chainwheel and sprockets on bicycles, and the gearwheels and flywheels in motorcycle, car and truck engines.

Motor vehicles have become essential to our way of life. Cars and motorcycles give us freedom to travel when and where we will. Heavier commercial vehicles supply the needs of industry and commerce. But so many vehicles now use the roads that traffic, especially in cities, is grinding to a halt; and the environment is being put at risk by the fumes their engines give out.

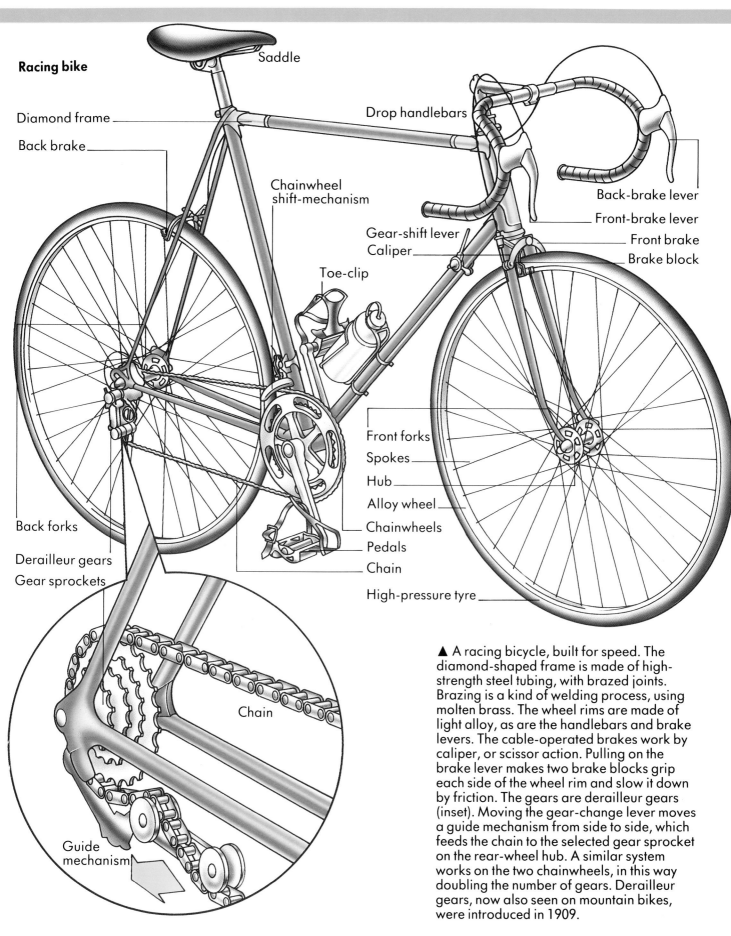

Racing bike

Saddle

Diamond frame

Back brake

Drop handlebars

Chainwheel shift-mechanism

Back-brake lever

Front-brake lever

Gear-shift lever

Front brake

Caliper

Brake block

Toe-clip

Front forks

Spokes

Hub

Alloy wheel

Back forks

Chainwheels

Derailleur gears

Pedals

Gear sprockets

Chain

High-pressure tyre

Chain

Guide mechanism

▲ A racing bicycle, built for speed. The diamond-shaped frame is made of high-strength steel tubing, with brazed joints. Brazing is a kind of welding process, using molten brass. The wheel rims are made of light alloy, as are the handlebars and brake levers. The cable-operated brakes work by caliper, or scissor action. Pulling on the brake lever makes two brake blocks grip each side of the wheel rim and slow it down by friction. The gears are derailleur gears (inset). Moving the gear-change lever moves a guide mechanism from side to side, which feeds the chain to the selected gear sprocket on the rear-wheel hub. A similar system works on the two chainwheels, in this way doubling the number of gears. Derailleur gears, now also seen on mountain bikes, were introduced in 1909.

The motorcycle

The motorcycle, or motorbike, combines features of both the bicycle and the car. The motorbike frame is broadly similar to that of a bike, with the front wheel held in forks, which are turned by the handlebars for steering. The drive from the engine is usually by chain to a sprocket on the rear wheel. A few models, however, have shaft drive, like a car.

The standard motorbike engine is a petrol engine that works on the four-stroke cycle like a car engine. Smaller bikes may have a two-stroke engine, which is simpler in design and easier to maintain. In most bikes the engine is air-cooled: fins are fitted around the engine cylinders to give a greater area for cooling.

As in a car, engine power is controlled by a throttle and transmitted through a clutch and gearbox. The rider's right hand is used to work the twist-grip throttle, which increases or decreases engine speed by allowing more, or less, fuel mixture into the engine cylinders. The clutch is operated by a lever on the left handlebar; and the gearbox by a gear-shift pedal, also on the left.

A motorbike may have drum or disc brakes, or a combination of the two. Applying the brakes forces brake shoes or pads against a drum or disc attached to the wheel. The front brake is applied by means of a hand lever on the right handlebar, while the back brake is applied by means of a foot pedal, also on the right.

Anatomy of a superbike

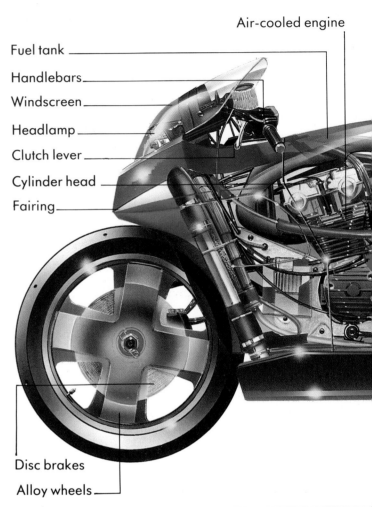

Fuel tank

Handlebars

Windscreen

Headlamp

Clutch lever

Cylinder head

Fairing

Air-cooled engine

Disc brakes

Alloy wheels

From steam bike to superbike

Ernest and Pierre Michaux (France) built a steam-powered motorbike in 1868. Gottlieb Daimler (Germany) developed the first petrol-driven machine in 1885, before turning his attention to motor cars. The modern form of the bike appeared in 1901, designed by Michael and Eugene Werner (France). It had the engine slung low down between the wheels and used a twist-grip throttle. Motorbike manufacturers sprung up throughout the world: for example, Norton, BSA and Matchless in Britain, Indian in the USA, and BMW in Germany. In the 1950s, motor scooters like the Vespa became popular. Since then Japanese makes, such as Honda, Suzuki, Yamaha and Kawasaki, have come to dominate the motorbike scene.

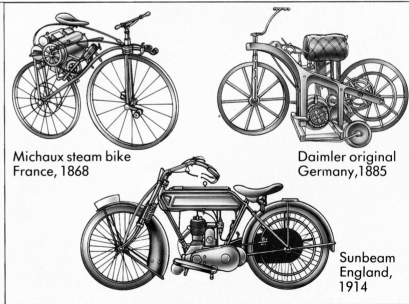

Michaux steam bike
France, 1868

Daimler original
Germany, 1885

Sunbeam
England,
1914

▼ The modern superbike has a powerful engine and breathtaking performance. On the track it can reach speeds up to 250 km/h. Its body is enveloped in a streamlined fairing to reduce air resistance. Disc brakes on both wheels ensure efficient and safe braking.

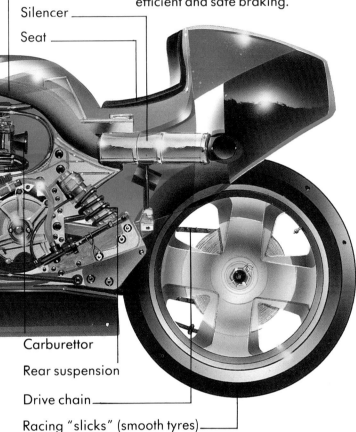

Exhaust pipe.

Silencer

Seat

Carburettor

Rear suspension

Drive chain

Racing "slicks" (smooth tyres)

▲ A motocross rider hustling his bike around a cross-country circuit. The machines used for motocross are rugged lightweight bikes, usually with a two-stroke engine. They are recognized by the high ground clearance of the body, which allows for excessive suspension movement over the rough, bumpy ground found on motocross circuits. They are fitted with tyres with a chunky tread, which give them good grip.

Indian
USA, 1916

Vespa scooter
Italy, 1950s

BMW
Germany, 1923

Yamaha
Japan, 1990

Car systems

The motor car is made up of as many as 14,000 different parts. These are assembled into larger components, which in turn are put together on the car production line. It is convenient to describe a car in terms of the systems that are needed to make it run. The engine unit forms one major system, and itself comprises many subsystems: fuel, ignition, lubrication, cooling, and so on. The engine power is carried to the driving wheels by the transmission.

In the conventional car layout shown here, the transmission is made up of the clutch, gearbox, propeller shaft and final drive. The clutch connects with the flywheel of the engine and acts like a switch to cut off power to the gearbox when the driver wants to change gear.

When the gear-change lever is moved, different sets of gearwheels in the gearbox mesh together to increase or decrease the speed of the output shaft. The propeller shaft carries the motion to the differential, which transfers the motion to the half-shafts that turn the wheels.

A variety of other systems help the driver control the car. He or she guides the car through the steering system and slows it down by the braking system. The foot brake works hydraulically and acts on all four wheels. The hand, or parking, brake works mechanically and acts on only the rear wheels. The car also has a suspension system to cushion passengers from jarring on bad road surfaces.

Steering system

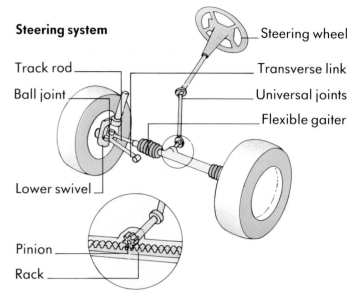

Track rod
Ball joint
Lower swivel
Pinion
Rack

Steering wheel
Transverse link
Universal joints
Flexible gaiter

▲ This kind of steering system is called rack-and-pinion. Turning the steering wheel rotates a small gearwheel, or pinion. This moves the toothed rack from side to side. The rack is linked to the front wheels by a pair of track rods.

▶ The major systems in this conventionally laid-out car are colour-coded. The engine, exhaust and fuel tanks, are mauve. So is the transmission system of clutch, gearbox, propeller shaft and final drive. The cooling system, which uses the radiator to cool hot water circulating from the engine, is blue. The braking system, which comprises disc brakes at the front and drum brakes at the rear, is pink. So is the suspension system, which incorporates springs and dampers, or shock absorbers. The electrical system, comprising the battery, coil, distributor, generator, lights, windscreen wipers and instruments, is orange.

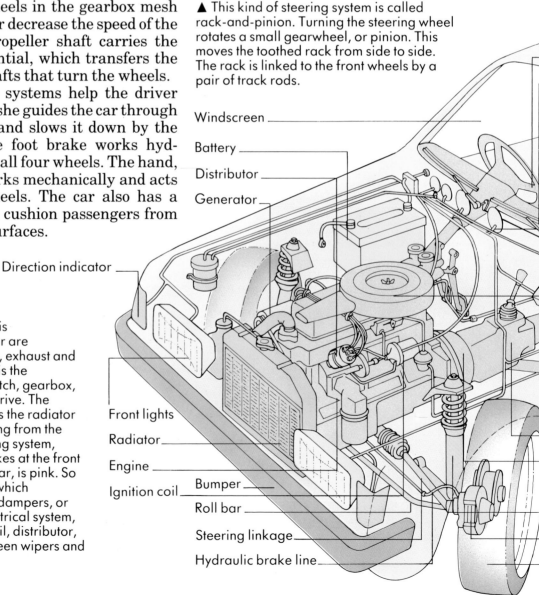

Direction indicator
Windscreen
Battery
Distributor
Generator
Front lights
Radiator
Engine
Ignition coil
Bumper
Roll bar
Steering linkage
Hydraulic brake line

► The clutch, gearbox and differential are key units in the transmission system. The clutch is mounted on the engine flywheel. The pressure plate forces the driven plate against the flywheel to transmit motion to the gearbox. It is sprung away from the driven plate when the clutch pedal is depressed. The purpose of the gearbox is to mesh different-sized gearwheels and change the speed of the output shaft. The design of the differential allows the two drive wheels to turn at different speeds and thus prevent skidding when travelling around corners. In drum brakes the linings are forced against a drum attached to the wheel and slow it down by friction. In disc brakes the brake pads are forced against a disc attached to the wheel.

Layout of a rear-wheel drive car

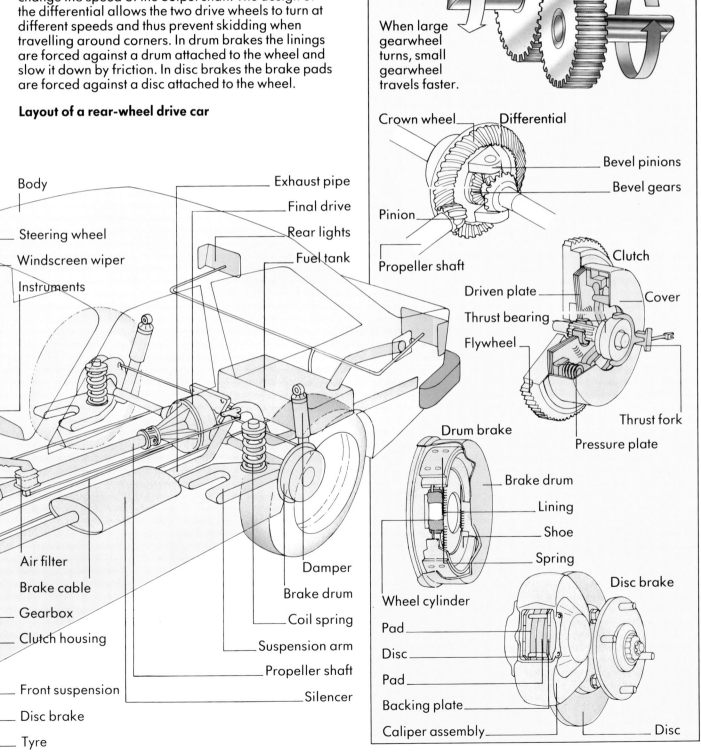

Transmission and brakes

Gearwheels

When large gearwheel turns, small gearwheel travels faster.

Crown wheel — Differential

Bevel pinions

Bevel gears

Pinion

Propeller shaft

Clutch

Driven plate

Cover

Thrust bearing

Flywheel

Thrust fork

Pressure plate

Drum brake

Brake drum

Lining

Shoe

Spring

Wheel cylinder

Pad

Disc

Pad

Backing plate

Caliper assembly

Disc brake

Disc

Body

Exhaust pipe

Final drive

Steering wheel

Rear lights

Windscreen wiper

Fuel tank

Instruments

Air filter

Damper

Brake cable

Brake drum

Gearbox

Coil spring

Clutch housing

Suspension arm

Propeller shaft

Front suspension

Silencer

Disc brake

Tyre

Car design

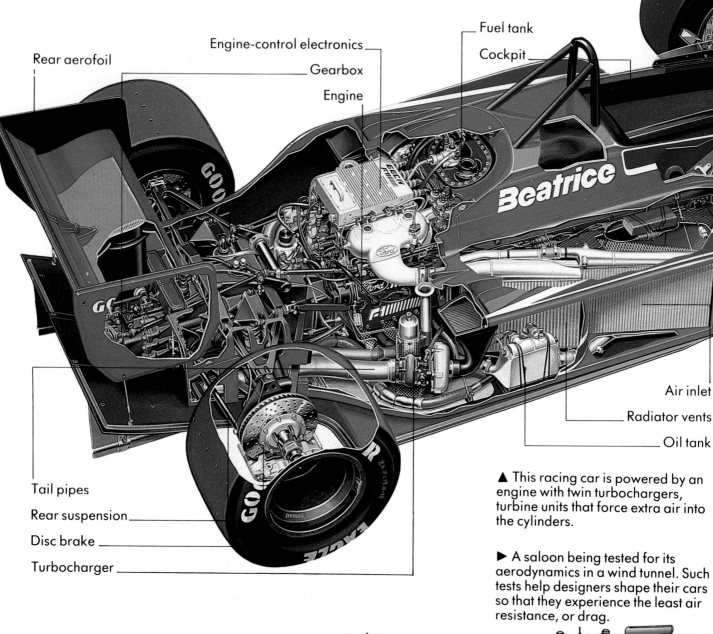

Rear aerofoil

Engine-control electronics

Gearbox

Engine

Fuel tank

Cockpit

Beatrice

Air inlet

Radiator vents

Oil tank

Tail pipes

Rear suspension

Disc brake

Turbocharger

▲ This racing car is powered by an engine with twin turbochargers, turbine units that force extra air into the cylinders.

► A saloon being tested for its aerodynamics in a wind tunnel. Such tests help designers shape their cars so that they experience the least air resistance, or drag.

From horseless carriage to supercar

In 1885 Gottlieb Daimler and Karl Benz in Germany fitted petrol engines to carriages to create the motor car. But the real motor-car revolution did not begin until 1908, when Henry Ford (USA) began mass-producing the Model T, or "Tin Lizzie". The revolution has continued with hardly a pause until the present day, when up to 50 million vehicles are produced worldwide every year.

Panhard-Levassor 1891

Benz three-wheeler Germany, 1885

Daimler Germany, 1885

Model T Ford USA, 1908

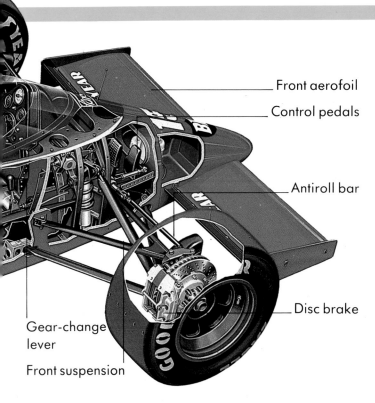

Front aerofoil

Control pedals

Antiroll bar

Disc brake

Gear-change lever

Front suspension

In a little over a century the motor car has evolved from a crude motorized horseless carriage into a sophisticated machine that is sleek, swift and comfortable. Car designs have changed markedly both inside and out as the years have gone by. Methods and materials of construction have altered. Engines have been progressively improved to be more economical and in recent years to reduce pollution. Diesel engines, once used only for trucks and buses, are now becoming increasingly popular for cars. They last longer, use cheaper fuel and give more economical motoring.

The most obvious feature of car design is body shape, which varies according to the type of car (saloon, hatchback, estate, coupé, convertible) and from manufacturer to manufacturer. Until the 1950s most car bodies were built by attaching body panels to a rigid frame, or chassis. Since then one-piece, or unitary, construction has been the norm. In this method the body is put together as a "shell" of shaped welded steel panels.

The car is a lethal weapon if it is in the hands of a bad driver or gets out of control. Car designers bear this in mind, and incorporate as many safety features as possible. They try to build the passenger compartment as a "safety cage" so that it is rigid enough to remain intact if the car crashes. They build "crumple zones" into the front and rear of the car to absorb much of the energy of an impact there. The doors may be reinforced with steel bars to prevent them caving in. Designers conduct crash tests to see how safe their designs are in practice.

Rolls-Royce
Britain, 1909

Volkswagen (VW)
Germany, 1938

Mini
Britain, 1960s

Lamborghini
Italy, 1970s

Citroen
France, 1934

Pontiac
USA, 1950s

Commercial vehicles

In business and industry the efficient transport of raw materials, products and goods is essential to keep production lines busy and customers happy. For such transport a wide variety of different vehicles is made, from light vans carrying stationery to heavy trucks carrying gravel from quarries. These load or freight carriers are known as commercial vehicles. In most countries, particularly the United States, trucking – carrying goods by truck – is big business.

Apart from freight carriers, there are many other kinds of commercial vehicles, for example buses and coaches to carry passengers over short and long distances; dustcarts to remove household rubbish; fire engines to fight fires; and rotating mixer trucks to carry wet cement to building sites.

Although these vehicles are very different in function and appearance, underneath they are basically similar. They are built up from an engine and chassis unit. They differ in the type of bodywork fitted to this unit. Usually the engine is a diesel, a rugged engine that runs on light oil. For extra power, it might be turbocharged, which means that extra air is forced into the engine cylinders by a turbopump.

Commercial vehicles that haul heavy loads require a gearbox with many more gears than a car so that they can cope with hills and different road conditions. A common arrangement is to

◀ One of the fire trucks of the town of Stonington, Maine, USA. It uses a special body fitted on a standard two-axle chassis. It carries escape ladders and has built-in pumps for forcing water through the fire hoses.

▶ Double-decker buses like this provide efficient, high-volume in-town transport. This is a Super Metrobus, now in service in Hong Kong with the Kowloon Motor Company. It is 12 m long and can carry up to 170 passengers. It has a turbocharged diesel engine and automatic transmission. The body is made of aluminium, and the passengers enjoy a comfortable ride because of the air suspension.

◄ An articulated truck passing through a small town in New England, USA. The tractor unit has three axles with four wheels on the rear two. It has the distinctive vertical silencer and exhaust pipe of the American truck. Like all heavy trucks, it has powerful air brakes, applied using air compressed by an engine-driven compressor.

▼ A dumper truck removes a load of rock from a quarry in northern England. It is a rugged, off-the-road vehicle with large, chunky tyres, and all four rear wheels are driven to give extra traction over rough terrain. The hydraulic ram that tips the body is clearly visible. Dumper trucks are also widely used in the construction of roads.

have two gearboxes in tandem: a main gearbox with up to six forwards gears and one reverse, and a "splitter" gearbox with two gears, making 14 gears in all.

The smaller commercial vehicles have just two wheel axles, like a car. Bigger ones may have as many as four axles to distribute their load more evenly. In these the front two axles are used for steering, and the rear two both drive axles.

Other vehicles are articulated. The power unit, or tractor, is separate from the load-carrying part, the semitrailer. The two parts connect via a coupling called a turntable, which allows them to swivel independently. This arrangement makes for easier manoeuvring and greater flexibility of operation: any tractor is able to haul any semitrailer.

On the rails

Spot facts

- *The standard gauge of railway lines – the distance the rails are apart – is 143.5 cm. The other most common gauge is the metre gauge.*

- *The heaviest and most powerful steam locomotives ever built were Union Pacific Railroad's "Big Boys". They weighed no less than 600 tonnes, had 16 driving wheels, and had the pull of 6,000 horses.*

- *Japan has the busiest railway system in the world, carrying over 20 million passengers daily.*

- *London has the world's oldest (from 1863) and most extensive underground railway network, with over 400 km of track, of which more than a third is in deep tunnels.*

- *A specially-prepared TGV broke the rail speed record with a run of 482 km/h in December 1989.*

▶ An electric locomotive hauls an express train through the scenic Rhine Valley in Germany. German Railways (Deutsche Bundesbahn) maintains one of the world's most efficient rail networks.

The coming of the railways in the early part of the 1800s brought about a revolution in transport. On the "iron road" people could travel long distances in comparative comfort, speed and safety for the first time. For more than a century steam locomotives hauled the trains, but more efficient electric and diesel locomotives have since taken over. The advantages of the railways for transport are that they can carry enormous loads, of passengers or freight, and make very efficient use of fuel. This is because there is minimal friction between the steel wheels of the trains and the steel track they run on.

134

The Railway Age

In 1804 Richard Trevithick put a steam engine on rails to create the first locomotive. But the Railway Age did not really begin until 1825. In that year the British engineer George Stephenson completed the first public steam railway, the Stockton and Darlington Railway. Five years later he built the Liverpool and Manchester Railway, and also the locomotive that hauled the first train on it, *Rocket*. Railway mania began to spread worldwide.

In the USA the railways opened up new territories for settlement: in 1869 the first transcontinental railway was completed, worked by American standard locomotives, with huge chimney and cowcatcher. By the 1930s in Britain streamlined locomotives like *Mallard* were setting speed records in excess of 200 km/h. In the USA in the 1940s monster locomotives like Union Pacific's "Big Boys" were being built, exceeding 600 tonnes in weight. But the days of steam were drawing to a close. Diesel and electric locomotives were beginning to take over.

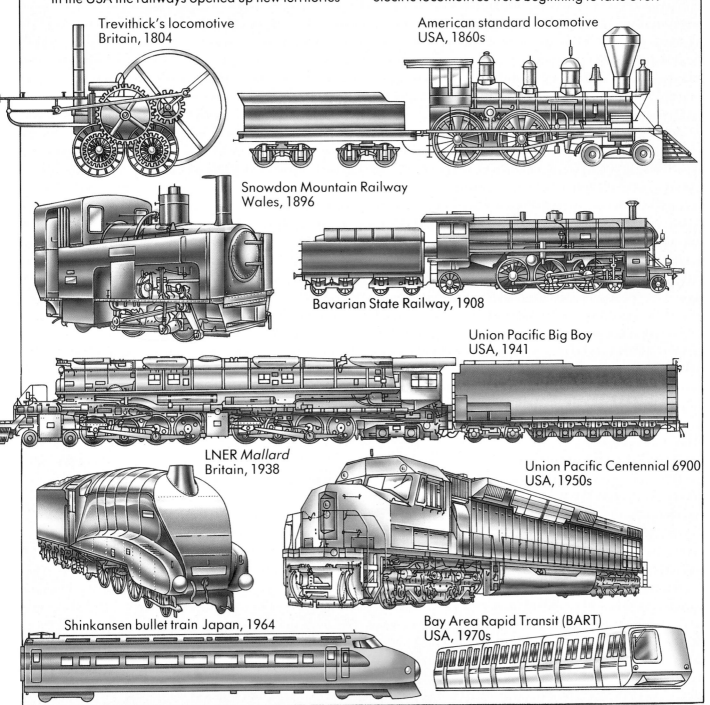

Trevithick's locomotive
Britain, 1804

American standard locomotive
USA, 1860s

Snowdon Mountain Railway
Wales, 1896

Bavarian State Railway, 1908

Union Pacific Big Boy
USA, 1941

LNER *Mallard*
Britain, 1938

Union Pacific Centennial 6900
USA, 1950s

Shinkansen bullet train Japan, 1964

Bay Area Rapid Transit (BART)
USA, 1970s

Locomotives

Two main kinds of locomotives haul modern trains: electric and diesel locomotives. Electric locomotives are the most common type in Europe; they were introduced into mainline service in Italy in the 1920s. Diesels are particularly popular in the United States. Streamlined diesel passenger trains were introduced there in the mid-1930s, shortly after Germany's *Flying Hamburger* train (1932) had pioneered diesel passenger travel.

Diesels use the same kind of piston engine as heavy trucks, and burn light oil as fuel. They are known as compression-ignition engines. This is because the fuel ignites when it is injected into hot, highly compressed air in the cylinders. Apart from this, the engines operate in the same way as car petrol engines.

Engine power may be transmitted to a diesel locomotive's driving wheels in three main ways. In diesel-electric locomotives, the engine drives an electricity generator. Then the current produced is fed to electric motors, which turn the driving wheels. Diesel-hydraulic locomotives use a kind of liquid coupling between the engine and the wheels. Diesel-

Preserved steam

Steam locomotives have in most countries been relegated to the history books, although examples can still be seen working in Eastern Europe, South America and India. In other countries steam railways are kept alive by bands of dedicated enthusiasts. They rescue and repair abandoned locomotives and run them on preserved lines that they have bought.

Britain is at the forefront of steam-locomotive preservation, as befits the country where the railways were first developed. One of the star attractions at "living-steam" displays is the *Flying Scotsman* (right), a renowned Pacific Class (4-6-2) locomotive, which was built in 1923. In 1934 it became the first locomotive in the world to reach a speed of 100 miles an hour (160 km/h). In its 40-year working life, it travelled more than 3,300,000 km.

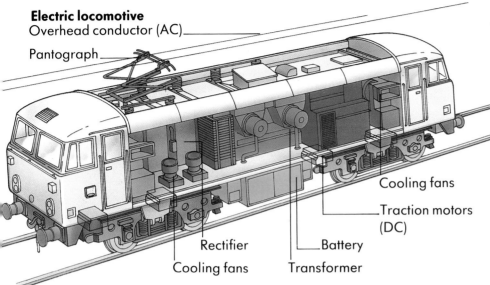

Electric locomotive
Overhead conductor (AC)
Pantograph
Cooling fans
Traction motors (DC)
Rectifier
Battery
Cooling fans
Transformer

Third-rail pick-up

Some electric locomotives pick up current, not from an overhead conductor, but from a third insulated rail laid alongside the ordinary track. They are fitted with a pick-up "shoe" that slides along the third rail. Some mainline and many underground railway systems use the third-rail system, operating on DC.

Brake
Motor control
Pick-up "shoe"
Traction motor

◀ Two diesel-electric locomotives head a freight train on the Union Pacific Railroad in the western USA. They are among the most powerful diesels in the world, with a power output of over 6,500 horsepower. As many as four diesel units may be used in the western USA to haul freight trains measuring several kilometres long.

▲ A mainline electric locomotive operating on a 25,000-volt alternating current (AC) supply. The pantograph picks up current from the overhead conductor and feeds it to a transformer, which reduces the voltage. Next, the current is converted to direct current (DC) by a rectifier and fed to the traction motors that drive the wheels.

mechanical locomotives, mainly used for shunting, have a mechanical gearbox, rather like a truck.

Electric locomotives are very clean, quiet, pollution-free and highly efficient. Their main drawback is that they can run only on special track, whereas diesels can run on any track. On most electrified track, the locomotives pick up electricity from an overhead line, or conductor. They carry on top a sprung arm, or pantograph, which makes contact with the conductor. Other electric locomotives pick up current from a third rail, laid alongside the usual track.

Most overhead conductor systems operate at a voltage of 25,000 volts AC (alternating current). On board the locomotive, this voltage is first reduced by a transformer, and then converted to DC (direct current) by a device called a rectifier. The current is fed to DC motors, which drive the wheels.

There are also a few gas-turbine locomotives in some countries, notably the United States, Canada, Russia and France. They are propelled by a turbine spun by burning gas.

High-speed trains

In most countries the railways were built 150 or more years ago. Over the years the track has been improved by using more rigid concrete sleepers, or supports, and by the use of continuous-welded rail, often several kilometres long. The use of such rails makes for a smoother and quieter ride. It eliminates the relentless "clickety-click" noise and vibration produced when wheels ran over the regular joints between the once-common shorter lengths of rail.

Nevertheless, because of curves and slight gradients on most tracks, trains cannot usually average a high speed, no matter how powerful the locomotive used. To achieve consistent high speeds, special tracks need to be built. And this is what has happened in Japan and France. They are built as straight and as flat as possible and have no crossing points with old tracks and few, if any, signals.

Train à Grande Vitesse (TGV)

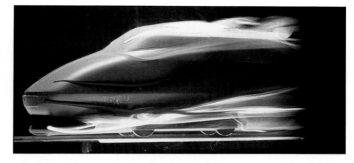

▲▼ The severely streamlined shape of the TGV (below) was designed after extensive testing of models in a high-speed wind tunnel (above).

Key

1 Collision protection
2 Brake gear
3 Driver's cab
4 Cooling air vents
5 Traction motors
6 Driver's cab air-conditioning
7 Battery compartments
8 Main transformer
9 Suspension
10 Main compressor
11 Pantograph (1,500V DC)
12 Pantograph (25,000V AC)
13 Overhead wires
14 Rectifiers
15 Baggage compartment
16 Automatic exterior doors
17 Passenger compartment
18 Inter-car bogie

▶ One of the famous Japanese bullet trains on the Tohoku section of the Shinkansen network, which opened in 1982. The total network now extends for nearly 2,000 km.

Japan pioneered such rail "expressways" as long ago as 1964, when its Shinkansen, or new trunk line, opened. The first section, the New Tokaido line, ran between Tokyo and Osaka. It was operated by the streamlined, futuristic-looking "bullet trains". These electric trains achieved *average* speeds of over 160 km/h. The network has since expanded west to Hakata and north to Niigata and Morioka.

The Shinkansen has now been surpassed as far as speed is concerned by France's TGV network. TGV stands for Trains à Grande Vitesse, meaning high-speed trains. And it is an apt name, for they regularly travel at speeds up to 270 km/h. The first stretch of newly-built track opened in 1981, between Paris and Lyons, in south-east France. The trains regularly make this 400-km journey in about two hours. Like the bullet trains, the TGVs form an integrated trainset of power units and passenger cars, the whole of which is streamlined to reduce air resistance.

Special railways

Ordinary railway track must be built as flat as possible. If it slopes too much, the wheels cannot grip the rails properly and start to slip. Yet in the European Alps and other mountainous parts of the world, some railways climb gradients as steep as 50 per cent, or 1 in 2: for every 2 m travelled the track rises 1 m.

One type is the rack railway, which uses a rack-and-pinion system to climb. There is a rack, or toothed rail laid between the two rails of the usual track. The passenger cars have two pinions, or cogwheels, fitted underneath, which engage the teeth of the rack. Other mountain railways are hauled by cable using a powerful winch. They often work in pairs on a funicular system: one going up as the other goes down.

Underground railways, or subways, are the quickest method of transport in many cities. Those in London, Paris and New York are among the oldest underground systems; those in Washington and Hong Kong are among the newest and are highly automated. Most underground trains are electric and pick up their electricity from a third rail. The power units are incorporated in the passenger cars.

San Francisco's cable-cars

The famous cable-cars of San Francisco, in California, USA, operate on a unique drive system. In a slot beneath the track runs a moving cable. Each cable-car has a hand-operated clutch, which grips the cable when applied, and the car is hauled along.

Railway operation

Passenger travel is served by the fastest trains and is the most obvious side of railway operation. But it is not the most profitable. On most railways freight traffic provides most of the income. And for moving goods in bulk, long distance, the railways are unbeatable.

All manner of freight is carried by rail in a variety of freight cars: coal in open wagons; oil in tanker cars; fruit in refrigerated cars; and trailers and containers on flat cars.

Containerization – transporting goods in standard-sized containers – is becoming increasingly common. Specialist handling equipment operates at container terminals at truck depots, stations and ports to transfer containers between truck, railway and ship.

On busy railway routes, especially in the morning and evening rush hours, trains follow one another along the same track with only minutes between them. This creates very real safety problems, and so over the years various traffic control systems have evolved. These days regional computerized control centres plan and monitor traffic movements of hundreds of trains over hundreds of kilometres of track. They control the signals and the switching of trains to different lines. They follow the progress of each train on a miniature track layout. Positional information is provided by electrical relays along the track, known as track circuits.

▲ The operations control centre in Paris for the TGV (Train à Grande Vitesse) network of French high-speed trains, which run from Paris to Lyons. Controllers are in continuous radio contact with every train.

◀ A passenger car on the Mount Pilatus rack railway in Switzerland, which opened in 1889. It is one of the steepest mountain railways in the world, climbing gradients of 1 in 2 on some stretches.

▶ Handling a container at a railway container terminal in Britain. Lifting gear from an overhead gantry grips and lifts the container and moves it from train to truck-trailer or ship, or the other way round.

On the water

- The world's largest dry cargo ship, the Norwegian vessel Berge Stahl, measures 342 m long – over 40 m longer than the height of the Eiffel Tower.

- A nuclear-powered submarine can remain underwater for months at a time and needs refuelling only about once every ten years.

- The inventor of the telephone, Alexander Graham Bell, in 1918 designed a successful hydrofoil boat, the HD4, which could travel at 60 knots (110 km/h).

- A supertanker, laden with perhaps half a million tonnes of crude oil, can take 2-3 km to stop.

► With colourful spinnakers deployed, ocean-going yachts run before the wind. Sails provided the main form of propulsion on water until about a century and a half ago.

The wooden sailing ship vies with the horse-drawn wagon as being the oldest form of transport. Throughout history ships have been used for commerce (merchant ships) and for fighting (naval ships), as they are still. A wide variety of ships sail the seas today, from tiny fishing boats and tugs to monster tankers, some the size of three football pitches laid end to end. Under the surface lurk the most formidable of fighting vessels, nuclear submarines, powered by nuclear reactors and armed with nuclear missiles. Skimming over the surface are novel craft such as hydrofoils and hovercraft. With their hull out of the water, these surface skimmers have speeded up water transport dramatically.

Development of ships

From sail to steam

The Ancient Egyptians were sailing square-sailed boats on the River Nile at least 6,000 years ago. By 1000 BC the Greeks were sailing to war in fast ships called triremes, which had three banks of oars. The Vikings used sail and oars in their longships of the AD 900s. By the 1400s ships had adopted the stern rudder and the triangular lateen sail, which enabled them to sail close to the wind. Three-masted vessels carrying a combination of square and lateen sails became standard. They included the *Golden Hind*, ship of the English explorer and adventurer Francis Drake. By the mid-1800s the speed merchants of the oceans were the great tea clippers. But the Age of Sail was all but over. Steamships began to appear. The *Great Britain* pioneered modern ship design, with its iron hull and screw propeller, although it still sported sails as well as a steam engine. Charles Parson's *Turbinia* introduced steam-turbine propulsion, now used by many large vessels, such as the world's longest liner *Norway*, which measures over 315m.

Nile boat
Ancient Egypt

Trireme
Ancient Greece

The Ark Royal
England, 1587

Viking longship
AD 900s

Clipper
England/USA
1850s

Great Britain
England, 1845

United States
USA, 1952

Turbinia
England, 1897

Ship shapes

Trading ships have sailed the seas for more than 5,000 years. And they are nearly as important today as then, despite competition from the air. The ship may be very much slower than the plane, but it can carry very much greater amounts of cargo very much more cheaply.

For carrying passengers over a long distance, the ship now takes second place. The most common working passenger ships are ferries on short sea crossings. The days of the transatlantic passenger liner are long gone. Most liners today operate as cruise ships, transporting tourists to sunny and exotic locations, such as the Mediterranean and the Caribbean.

Ships vary widely in design, according to their use. Passenger liners are easily recognized by their extensive superstructure – the part of the hull above the main deck level. They can be much like floating towns, with swimming pools, shops, restaurants and theatres.

In contrast cargo ships, or freighters, have little superstructure. There is usually just the

▲ A freighter fully laden with standard-sized containers, which are loaded and unloaded at special container terminals. The container method of carrying freight allows goods of all types, shapes and sizes to be transported with the minimum handling problems.

Screw propeller

Practically all ships are driven by propeller, often called a screw because it in effect screws itself through the water. The Swedish-born engineer John Ericcson developed the propeller in 1836. It soon replaced the paddle wheel.

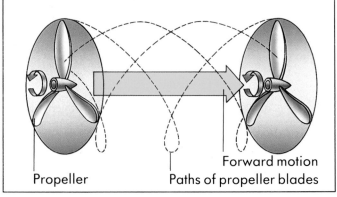

Propeller Paths of propeller blades Forward motion

navigation bridge and the smoke stack, below which are the boilers and engines. Most of the hull is taken up with the cargo space. In tankers, for example, this space consists of several separate tanks. They have to be separate to prevent the massive surging of the oil when the ship pitches and rolls. This could readily cause the ship to capsize.

Freighters carrying mixed cargo can usually be recognized by the rows of mastlike derricks, or cranes, on deck. More specialist freighters include Ro-Ro (roll-on, roll-off) vessels. They carry trucks and trailers, which are driven on at one port and off at another. Container ships transport goods in standard-size boxes, or containers, which are loaded and unloaded by special gantries. These transfer the containers from and to flat-top truck-trailers or railway cars. A similar idea is behind the LASH (lighter aboard ship) system. Goods are packed in standard-sized barges, which are then floated out to a LASH vessel and hauled aboard.

▲ A flotilla of tugs tows out a massive Condeep oil-production platform from the construction site towards the North Sea oilfields. Tugs are the tough workhorses of the oceans. Among their other jobs, they tow barges, help manoeuvre ocean liners into port and salvage stricken vessels.

▶ An Iranian oil tanker in the Persian Gulf making for a terminal to pick up a cargo of crude oil, or petroleum. Known as VLCCs (very large crude carriers), such vessels can carry hundreds of thousands of tonnes of oil.

Surface skimmers

Hydroplanes and hydrofoils

Even the fastest ordinary ships cannot travel much faster than about 35 knots, or 65 km/h. This is because they expend most of their engine power in overcoming the drag, or resistance of the water, on their hull.

To achieve high speeds, a vessel must somehow raise as much of its hull as possible out of the water. A racing speedboat, or powerboat, achieves this by hydroplaning. As its powerful engine drives it forwards, its bows (front) begin to rise out of the water. Most of the rest of its hull follows until only the stern (rear) and the propellers underneath remain submerged. The boat just skims the water.

Hydroplaning is not practical with larger, heavier boats. But they too can skim the surface if they are fitted with underwater "wings". These "wings", called hydrofoils, develop lift when moving through the water in much the same way that aircraft wings (aerofoils) develop lift when moving though the air. Hydrofoil boats have two sets of hydrofoils fitted fore and aft beneath the hull.

When stationary, and at low speeds, the hull rests in the water, as with an ordinary boat. But as speed increases, the foils start lifting upwards. They continue to lift until they raise the hull clear of the water, and only the struts connecting the foils to the hull and the propeller shaft are still in the water. Now almost free from drag, the boat can accelerate to speeds

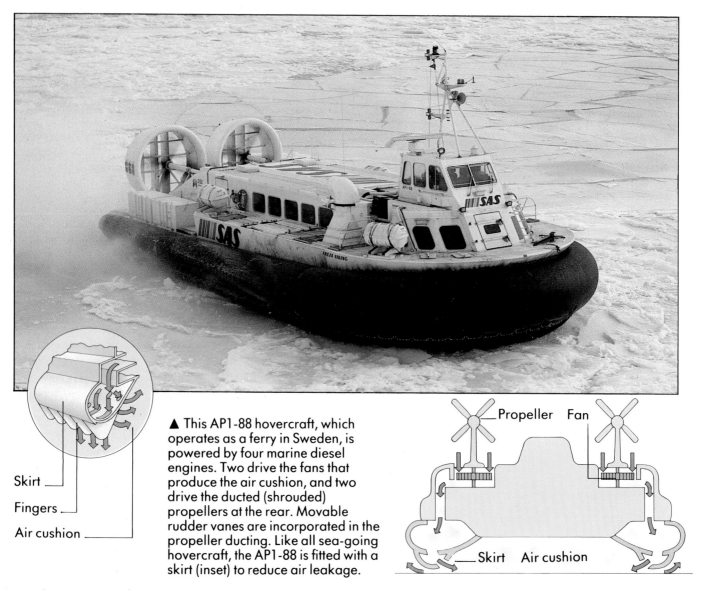

Skirt

Fingers

Air cushion

▲ This AP1-88 hovercraft, which operates as a ferry in Sweden, is powered by four marine diesel engines. Two drive the fans that produce the air cushion, and two drive the ducted (shrouded) propellers at the rear. Movable rudder vanes are incorporated in the propeller ducting. Like all sea-going hovercraft, the AP1-88 is fitted with a skirt (inset) to reduce air leakage.

Propeller   Fan

Skirt   Air cushion

146

Hydrofoils

Two main kinds of hydrofoils are fitted to boats. Surface-piercing foils, of which two types are shown below, extend out of the water. Fully submerged foils remain beneath the surface. The V-foil is most widely used but is not suitable for rough waters. The Jetfoil (right) is fitted with fully submerged foils.

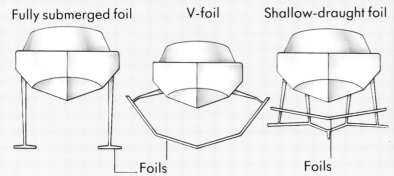

Fully submerged foil   V-foil   Shallow-draught foil

Foils   Foils

▲ A rear view of an SRN4 hovercraft ferry, showing vehicles unloading and passengers disembarking at a hoverport on the English Channel. The SRN4 is nearly 57 m long and can carry over 400 passengers and 60 cars. The picture shows the "pushing" propellers mounted on pylons, and the twin rudders. Also visible are three of the four engine exhaust nozzles.

◄ The gas-turbine engines of the SRN4 each drive a propeller geared to a fan. The fan sucks in air and forces it through a skirt to create an air cushion.

maybe as high as 60 knots (110 km/h).

Hydrofoil vessels are used, particularly in Europe, for swift river transport and as ferries on short sea crossings. The most sophisticated hydrofoils are the Jetfoils built by the aircraft manufacturers Boeing. They have gas-turbine engines and are notable by being driven by twin water jets.

Hovercraft

On some routes hydrofoil boats have a rival surface skimmer, the hovercraft. This is a kind of air-cushion vehicle, so called because it travels along on a high-pressure bubble, or "cushion" of air. The hovercraft can travel over virtually any surface: solid ground, swamp or water. But it is for water transport that it has made its mark.

Outstanding are the SRN4 hovercraft ferries that ply the English Channel, which can reach 65 knots (120 km/h). They have four gas-turbine engines, which drive powerful fans to force air underneath to form a cushion. A flexible "skirt" reduces air leakage. The hull is lifted clear of the water, largely eliminating drag. The engines also drive four large air propellers to provide propulsion. The propellers face backwards, operating in a "pushing" mode. Rudders at the rear, in the slipstream of the propellers, provide directional control.

Submarines and submersibles

Today, the most formidable fighting ships are not surface vessels but nuclear submarines. These craft can lurk undetected in any of the Earth's vast oceans, and can carry enough nuclear missiles to devastate more than a hundred cities.

Submarines submerge by flooding ballast tanks with water, and surface by emptying them using compressed air. Like ordinary ships, they use a propeller for propulsion. It is driven in conventional submarines by diesel engine on the surface, and by electric motor underwater. Nuclear submarines are driven by steam turbines. The steam is produced in a generator heated by a compact pressurized-water nuclear reactor.

A submarine is steered through the water by means of a vertical rudder and horizontal fins, called diving planes or hydroplanes. There is one set of planes in the cross-shaped "tail" at the stern, and one on each side near the bows or on the projecting conning tower, or sail. The sail acts as a navigation bridge on the surface. It also houses one or more periscopes and a snorkel tube. The periscopes enable the crew to see above the surface while still remaining submerged. The snorkel enables them to take in air while still submerged.

Nuclear-powered submarines are large vessels: those belonging to Russia's Typhoon class are 170 m long and displace 25,000 tonnes. The submarine craft in civilian use, called submersibles, are very much smaller. They are now used widely for ocean research and for supporting offshore oil production work. Many are so-called lockout submersibles. They have a separate diving compartment, which can be pressurized to match that of the sea.

▶ The US submarine *Will Rogers* is a nuclear-powered, missile-carrying vessel, able to travel underwater for months at a time and go for 600,000 km or more without refuelling. It has advanced sonar and radar systems. The US pioneered nuclear propulsion for submarines with *Nautilus* in 1954.

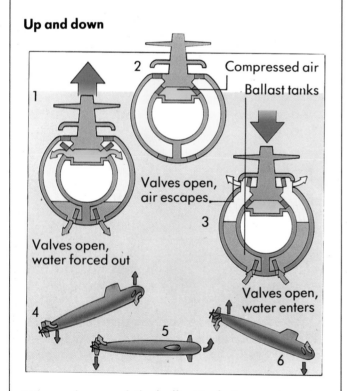

Up and down

Compressed air

Ballast tanks

Valves open, air escapes

Valves open, water forced out

Valves open, water enters

When submerged, the ballast tanks in a submarine contain water. To surface (1), the water is blown out by compressed air. On the surface, the tanks are empty (2). To submerge, water is let into the tanks (3). Moving the diving planes fore and aft puts the nose up (4) or down (6). Moving the rudder achieves sideways control (5).

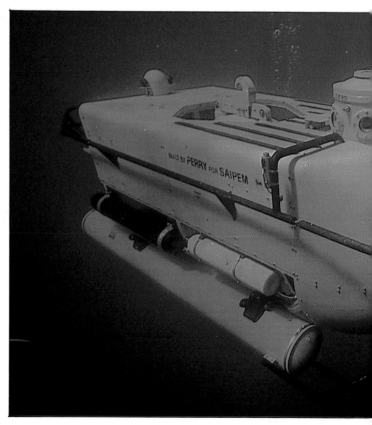

Lighter than air

When a pocket of air is heated, it expands. It becomes lighter, or less dense, than the surrounding air. It therefore rises. This simple scientific principle was behind the design of the first aircraft. They were fabric and paper bags containing hot air. Beneath the open neck of the bag a fire was suspended to keep the air warm.

Two French brothers, Jacques and Etienne Montgolfier, launched the first such balloon at Annonay in June 1783. The hot-air balloon was born. Four months later the noted French physicist J.A.C. Charles launched a quite different design, filling a bag with hydrogen, the lightest of all gases. By the end of 1783 both kinds of balloons were carrying human passengers. Air transport had begun.

In 1852 a French engineer, Henri Giffard, fitted a steam engine to a balloon to create the first dirigible (steerable balloon), or airship. But the airship did not become a practical form of transport until 1900, when a German Count, Ferdinand von Zeppelin, built the first rigid craft. Like his later designs, it had an aluminium frame covered in fabric. It used bags filled with hydrogen gas to provide lift.

In World War 1 (1914-18) Zeppelins carried out the world's first air raids, dropping bombs on London. In 1919 the British airship *R34* made the first two-way crossing of the Atlantic. Ten years later the *Graf Zeppelin* circumnavigated the world in 21 days. It appeared that there was a great future for the airship. But it was not to be. A series of accidents in the 1930s, including the loss of the 247-m long Zeppelin *Hindenburg* in 1937, signalled the end of the airship era.

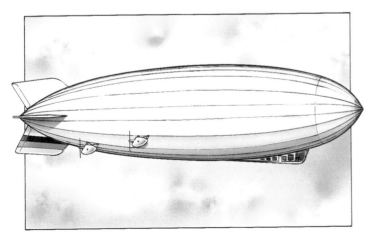

▲ A modern hot-air balloon. Hot-air ballooning has become a popular sport in recent years. The open fire of the Montgolfier balloon has been replaced by a burner that runs on bottled gas (propane).

◀▲ The German rigid airship *Graf Zeppelin* (left), launched in 1928. Between 1933 and 1937 it operated a transatlantic service between Germany and South America. The modern airship *Europa* (above) is a non-rigid ship, or blimp. It is filled with the non-flammable gas helium, not hydrogen.

Heavier than air

Every kind of aircraft must lift itself into the air against gravity, and then thrust itself forwards against the drag, or resistance of the air. The thrust may be provided by a propeller, or airscrew, or by a jet of hot gases.

The airship is an aircraft that is filled with gas to make it lighter than air. The aeroplane, the commonest kind of aircraft, is a heavier-than-air machine. It lifts itself into the air by means of its wings, much like a bird does. It uses the principles of aerodynamics, the science of flowing gases.

The wings of a plane have an aerofoil shape: broad at the front and sharp at the rear; flat underneath and curved on top. When air moves past an aerofoil wing, it travels faster over the upper curved surface than it does under the lower flat one.

As a result, the pressure above the wing becomes lower than the pressure below it. This makes the wing tend to lift upwards. And the faster the air flow, the greater is the lifting force. When a plane travels fast enough, the lifting force increases until it is greater than the plane's weight, and the plane flies.

▲ The flight deck of an Airbus 310 airliner, showing a bewildering array of instruments.

▼ The action of an aerofoil wing produces a pressure difference between top and bottom, which creates lift.

Aircraft control

Air flow around an aerofoil

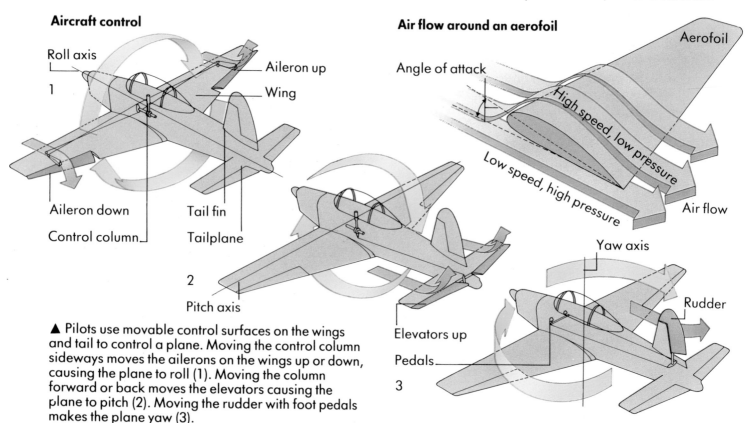

▲ Pilots use movable control surfaces on the wings and tail to control a plane. Moving the control column sideways moves the ailerons on the wings up or down, causing the plane to roll (1). Moving the column forward or back moves the elevators causing the plane to pitch (2). Moving the rudder with foot pedals makes the plane yaw (3).

From *Flyer* to Airbus

The English engineer George Cayley built the first heavier-than-air flying machine, a glider, in 1852. In the 1890s the Wright brothers (US) took up gliding and in 1903 attempted powered flight. On 17 December that year they succeeded, with their *Flyer*, a biplane. In 1909 Louis Blériot (France) crossed the English Channel in a monoplane. Planes like the Sopwith Camel (Britain) played a decisive role in World War 1 (1914-18). Speed records began to tumble: in the early 1930s the Supermarine S6B (Britain) became the first plane to exceed 400 miles an hour (640 km/h). By the late 1930s scheduled services had begun, operated by planes like the Douglas DC3 Dakota (US). Developments abounded during World War 2 (1939-45), including the first jet plane, the Heinkel-178 (Germany). Jet airline travel began with the Comet (Britain) in 1952. In 1969 the first jumbo jet, the Boeing 747 (US), made its maiden flight. Among today's most advanced airliners is Europe's "fly-by-wire" A320 Airbus.

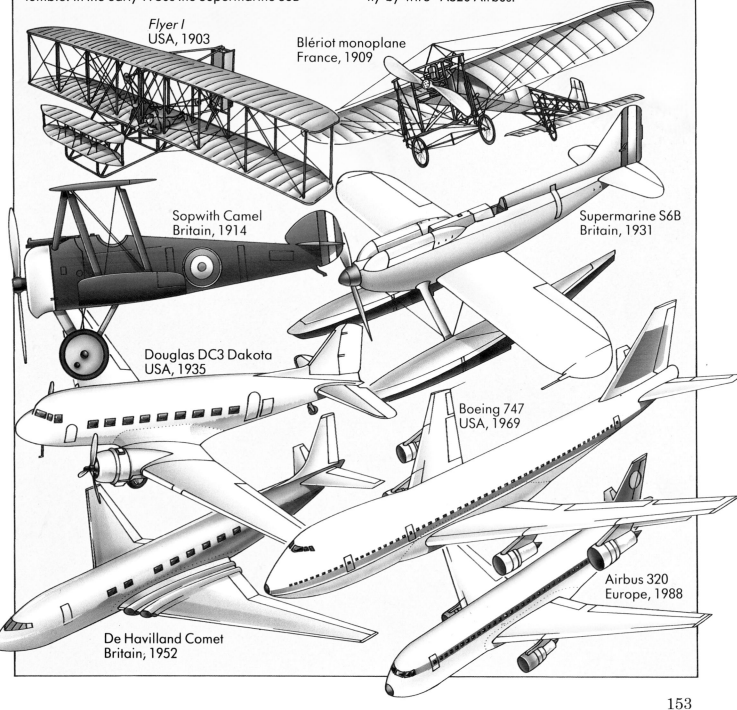

Flyer I
USA, 1903

Blériot monoplane
France, 1909

Sopwith Camel
Britain, 1914

Supermarine S6B
Britain, 1931

Douglas DC3 Dakota
USA, 1935

Boeing 747
USA, 1969

Airbus 320
Europe, 1988

De Havilland Comet
Britain; 1952

Aircraft design

Aircraft of all shapes and sizes fly the air routes of the world. They are differently designed to carry out their different roles. For example, a slow, heavy transport plane like the Super Guppy needs quite a different design from that of an agile, speedy attack aircraft like the F-16.

The Super Guppy has a bulky body. Its wings are long and thick, and stick straight out. It is propeller-driven, with a top speed of below 400 km/h. The F-16 by contrast has a sleek, narrow body, with a pointed nose. Its wings are short, thin and are swept back at an angle. It has a top speed of more than 2,000 km/h.

Long, thick, straight wings are typical of slow transport aircraft. These aircraft also use

▲ The unique aircraft *Voyager*, in which Dick Rutan and Jeana Yeager circumnavigated the world without refuelling from 14-23 December 1986. Its wingspan is 33.8 m.

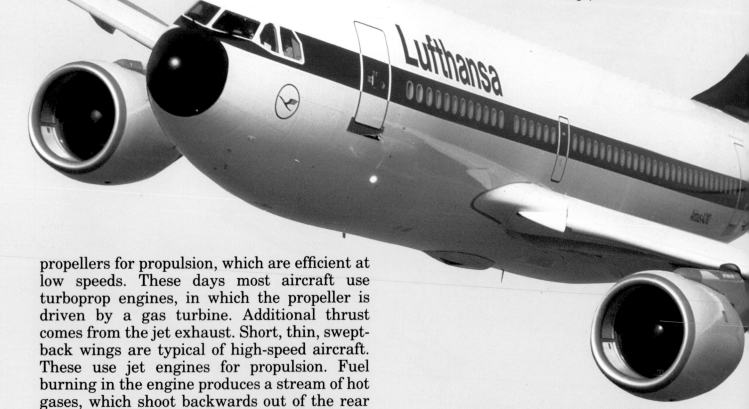

propellers for propulsion, which are efficient at low speeds. These days most aircraft use turboprop engines, in which the propeller is driven by a gas turbine. Additional thrust comes from the jet exhaust. Short, thin, swept-back wings are typical of high-speed aircraft. These use jet engines for propulsion. Fuel burning in the engine produces a stream of hot gases, which shoot backwards out of the rear nozzle, and propel the aircraft forwards by the force of reaction.

The design of airliners lies between these two extremes. They have quite a large fuselage to accommodate passengers and quite thick wings, which are swept back to some extent. Their two, three or four engines are jets, of a type known as a turbofan. These have a huge

▲ An Airbus 310 airliner of Lufthansa, Germany's national airline. The airliner is one of a successful series built by Airbus Industrie, a consortium of French, German, British and Spanish aerospace companies. It has a wingspan of nearly 44 m and a length of nearly 47 m. Its two turbofan engines give it a cruising speed at high altitude of nearly 900 km/h.

VTOL craft

A fully-laden Boeing 747 jumbo jet can weigh up to 350 tonnes. During take-off, it must travel for up to 3 km on a prepared runway before its engines have accelerated it to take-off speed. The need for a long runway for take-off, and landing too, is a disadvantage with ordinary aircraft. It is a particular drawback for the military, who need to be able to transport troops and equipment quickly in areas where there may be no runways at all.

It is at such times that they turn to vertical take-off and landing (VTOL) aircraft. The most common one is the helicopter. A more conventional-looking aircraft with VTOL capability is the Harrier "jump jet".

The helicopter is a nearly perfect flying machine, able to move in any direction in the air and hover like a hummingbird. Whereas aeroplanes have a fixed wing, helicopters have a rotary wing. This wing provides both the lifting

force and the propulsion for flying. The main drawback of the helicopter is its low operating speed, up to only about 300 km/h. It cannot travel much faster because otherwise the tips of the rotor blades would approach sonic speed, and the lift on the rotor would fail.

The blades of the helicopter rotor develop lift in much the same way as an ordinary wing, because they have the same aerofoil cross-section. To fly a helicopter, the pilot alters the amount of lift on the blades and varies the direction of the lift to propel it in any direction. He or she uses two main flight controls to achieve this: the collective-pitch lever and the cyclic-pitch lever.

To lift off the ground, the rotor blades are rotated and the pilot increases their pitch – the angle at which they hit the air – by moving the collective-pitch lever. This increases the lift of all the blades equally, and the helicopter takes off vertically. To travel forwards, backwards or sideways, the pilot operates the cyclic-pitch lever. This varies the pitch of each of the blades so that the resultant lifting force "pulls" the helicopter in the desired direction.

▼ The fixed-wing Harrier, developed in Britain, is a VTOL craft that works by what is called vectored (directional) thrust. Swivelling nozzles direct the jet stream from its engine downwards for vertical flight and backwards for horizontal flight.

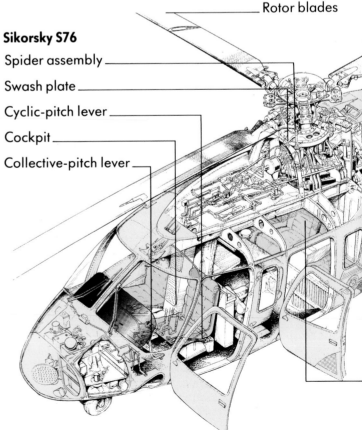

Rotor blades

Sikorsky S76

Spider assembly

Swash plate

Cyclic-pitch lever

Cockpit

Collective-pitch lever

▲ A Boeing Chinook helicopter working in the North Sea oilfields. This 30-seat craft has twin gas-turbine engines and twin rotors turning in opposite directions.

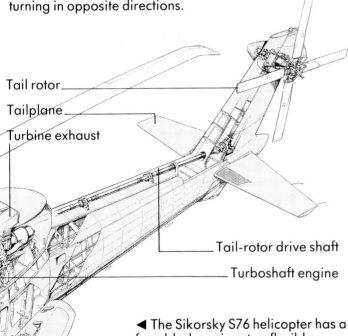

Tail rotor

Tailplane

Turbine exhaust

Tail-rotor drive shaft

Turboshaft engine

Landing gear

Passenger compartment

◀ The Sikorsky S76 helicopter has a four-blade main rotor, flexibly hinged at the hub. There is also a rotor spinning vertically at the tail. This is required in single-rotor craft like this to produce thrust to prevent the helicopter body spinning as the main rotor spins. This model has two gas-turbine engines, which turn the drive shafts to the rotors. Such engines are termed turboshafts. Linkages in the swash plate and spider alter the pitch of the blades.

▲ A helicopter hovers above a clearing amid a thick canopy of trees in the West African rain forest. It is air-lifting equipment being used in seismic surveying for oil. No fixed-wing craft could operate there. Helicopters have a host of commercial uses and are invaluable for sea and mountain rescue work. They can reach places inaccessible by any other means.

159

The printed word

Spot facts

• The Lawrence Livermore Laboratory in California, USA, houses the fastest printer in the world. It can print up to 3.6 million characters per minute: more than 3,000 times as many as the fastest typewriter. It could print The Bible in 65 seconds.

• The oldest printed book, a South Korean sutra, or scroll printed with wooden blocks, dates from about AD 700. This was more than 700 years earlier than Gutenberg's Bible, the first book to be mechanically printed, in 1454.

• The most modern digital phototypesetters can set up to 30 million characters per hour. By hand, 1,500 characters per hour can be set.

▶ Each copy of this book contains identical reproductions of the original pictures chosen to illustrate the subjects. As you read this book, other people all round the world can read the same book printed in their own language.

Every day we are all surrounded by print in newspapers, magazines, books, maps, posters, and countless other things. Printing allows words and pictures to be set down and copied at great speed. It has increased enormously the amount of information that we can now share about our world.

Many thousands of copies of this book have been printed. The words and pictures in the book have to be prepared for printing in a different way and then brought together for the print run to produce the finished book. Colour pictures are printed four times, each time in a different-coloured ink. Similar processes are used for all printed work. Many newspapers are now produced by the most advanced printing methods, using computers and lasers, and include colour pictures.

Typesetting

To print a page of a book, a copy of the words first has to be made out of metal. The completed metal letter will be covered in ink and pieces of paper pressed against it to transfer the ink to the page. The traditional way to produce the printed page is for a typesetter to select individual metal letters, each made as a mirror image of the letter to be printed. The case in which all the metal letters are stored has the capital letters in the upper part and the small letters in the lower part. This is why they are sometimes called upper-case and lower-case letters. The metal pieces are picked from the case and the typesetter arranges them in order, upside-down in a hand-held composing stick.

When the row of letters is complete, it is positioned as the bottom line on a frame called a galley. When the next line is complete, it is set above the previous line. The whole page is built up in this way. The finished galley contains an upside-down copy of the original page as it would appear in a mirror. When this is inked and used to print from, the printed page has the lines in the right order and all the letters the right way round.

Modern methods of typesetting avoid the lengthy process of arranging each metal letter individually. The author's writing can be typed into a word-processor and stored on a computer disc. The completed disc is then given to the compositor whose job is to translate the required page layout into computer commands on the disc. This will include details of the line length, spacing between lines, the typeface to be used, and the arrangement of the text on the page. Spaces will be left for pictures to be inserted later.

All these features of the finished page can be finalized on the computer screen without anything having to be printed. When the page of text is complete, the disc is transferred to a machine called a phototypesetter which produces a copy of the finalized text in the correct positions on transparent film. The shapes of all the letters are stored in the memory of the phototypesetter computer. A cathode-ray tube or a laser is used to transfer the shape of the letters required on to the film. This film is used in the manufacture of the metal printing plate, from which the paper page is printed.

▶ Traditional typesetting, which involves assembling lines of metal characters by hand. Many different styles of type are stored in the bench racks and drawers.

▼ Modern typesetting, which allows editors to change the text and decide on page layout on a computer screen. Phototypesetting machines can then set the type on to film at 1,000 lines per minute.

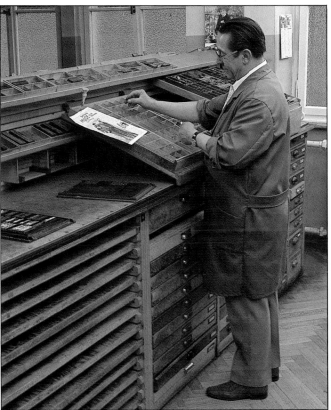

Origination

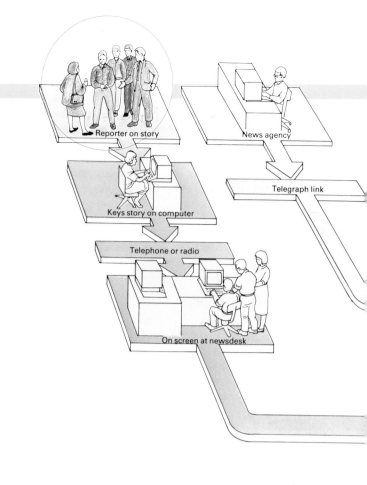

Origination is the copying of text or illustrations on to film, from which a printing plate can be made. Modern methods of phototypesetting can produce the film of the finished text directly. Otherwise the text is typeset in the traditional way and, once corrected, a photograph is then taken of a printed page.

Black-and-white photographs are reproduced every day in newspapers. They appear to be accurate copies of the original but in fact the printing process used for newspapers cannot print in grey, only in black. All the grey areas in the picture are made up of black dots on a white background (for light grey) or white dots on a black background (for dark grey). The original picture is converted to dots by photographing it through a screen containing many thousands of small holes. This screened photograph is used to make the printing plate.

Full-colour pictures, such as the illustrations in this book, are reproduced in a similar way. Colour pictures are printed using only black and three colours: yellow, cyan (a light blue), and magenta (a light purple), but the printed pictures you can see appear to contain all shades of colour. This is achieved by photographing the original three times through red, green and blue coloured filters. A screened copy of each photograph is then produced as for black-and-white photographs.

▲ Many newspapers use advanced technology to speed communications. By keeping words, illustrations and photographs stored in a computer, they can be used as required by journalists, editors, designers and printers. Journalists can type their stories directly into the computer. Photographs can also be scanned and entered into the computer. Editors design the layouts of the pages on screen. The finished design can then be typeset and made into film, from which printing plates are made.

▶ All printed full-colour pictures are made up of three coloured images, one in each of the three colours yellow, magenta and cyan. There is also a black image. These images, when put together, make a full-colour picture. First the full-colour original is photographed in black and white through a green filter. The resulting picture is re-photographed and screened into dots. From this a printing plate is made to be used with magenta-coloured ink. Similarly, using a red filter results in the printing plate for cyan ink, and a blue filter for yellow ink. To print a full-colour picture the paper must undergo four printing processes, one for each coloured ink and one for black ink. The resulting image will be a full-colour picture.

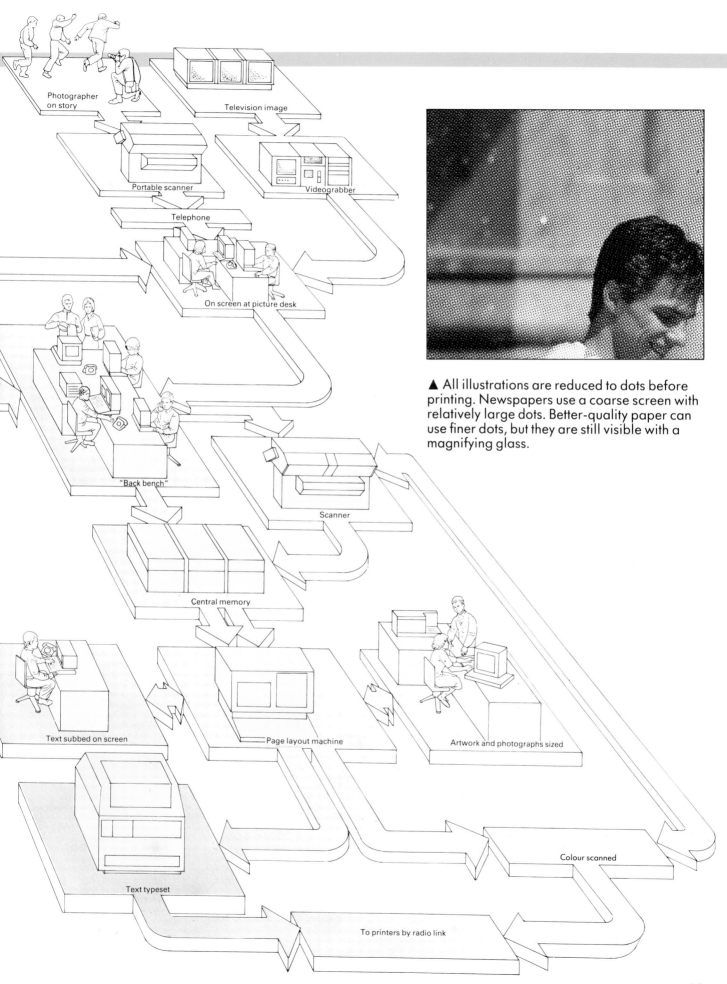

Photographer on story

Television image

Portable scanner

Videograbber

Telephone

On screen at picture desk

"Back bench"

Scanner

Central memory

Text subbed on screen

Page layout machine

Artwork and photographs sized

Text typeset

Colour scanned

To printers by radio link

▲ All illustrations are reduced to dots before printing. Newspapers use a coarse screen with relatively large dots. Better-quality paper can use finer dots, but they are still visible with a magnifying glass.

163

Printing methods

When the text has been typeset and pictures have been screened and made into final film, it is time to make the printing plate from which multiple copies will be made. All printing processes have to be designed to put ink on the paper in the places where it is required and nowhere else. There are three major printing processes, using different techniques to achieve this. These processes are called letterpress, gravure, and offset lithography. Newspapers are printed by offset lithography. Colour supplements are printed by gravure.

Although the printing plates for these processes are different, they are all made in a similar way. A metal plate is covered in a light-sensitive chemical. The transparent film of the page layout is brought close to the plate, and a bright light is shone through the film on to the plate. The chemical reacts where it is exposed to the light and in those areas it binds to the metal, acting as a protective layer on the surface. The metal is then exposed to a chemical process which either removes unprotected metal or changes its surface properties. In both cases the protective layer shields some areas of the metal from attack. The finished printing plates are wrapped around cylinders, which rotate as paper is pulled over them, transferring ink to the paper. A fourth method, silk screen printing, is often used for posters.

Gutenberg

Although books were being printed in China over a thousand years ago, the credit for introducing mass-production printing into Europe must go to Johannes Gutenberg. He lived in Mainz in West Germany in the 1400s. To print a page, the type was placed in a sliding tray and inked. The paper was placed on top, and the press was screwed down. This printing technique remained unchanged for three hundred years until rotating printing presses were invented.

▲ A web offset press prints a continuous roll of paper.

Impression cylinder

Paper sheet

Plate cylinder

Printed letter

Raised letter

Ink rollers

Letterpress

▲ The letterpress method is to etch away from the printing plate those areas which should not be printed. The raised areas are inked and the plate is rolled over the paper.

Ink consists of pigments, which form the colour, and additives, which control properties such as the stickiness or drying speed. A letterpress ink should be tacky and not flow very easily. In the gravure process, the ink must be thin enough to fill all the tiny cavities in the printing plate. It must also be sticky enough so that the ink is pulled out of the cavities when the paper is pressed against the plate. Lithography involves distinguishing between areas of the plate that receive ink and others which are wetted and repel ink. The ink must not run into the water but must fill all the unwetted areas. In all high-speed printing processes, ink needs to dry on the paper quickly.

Photocopying machine

The page to be copied is placed in the machine on a transparent surface. A rotating cylinder is first given a uniform electric charge. A bright light is moved beneath the original and an image of the page is projected on to the cylinder. Charge leaks away from those areas illuminated. A fine black powder is then sprinkled over the cylinder and is attracted to those areas still charged (the black areas). A blank piece of paper is moved over the cylinder, picking up the black powder. Heat then seals the powder on to the page.

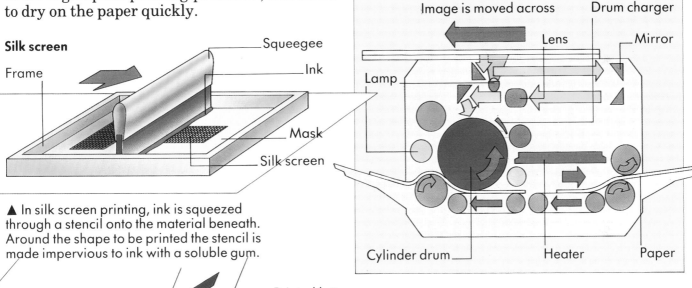

Silk screen

Frame — Squeegee — Ink — Mask — Silk screen

Image is moved across — Drum charger — Lens — Mirror — Lamp — Cylinder drum — Heater — Paper

▲ In silk screen printing, ink is squeezed through a stencil onto the material beneath. Around the shape to be printed the stencil is made impervious to ink with a soluble gum.

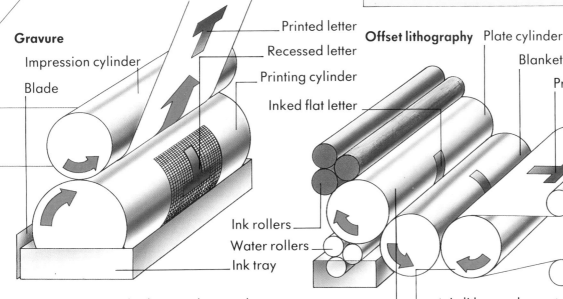

Gravure

Impression cylinder
Blade

Printed letter
Recessed letter
Printing cylinder
Inked flat letter

Ink rollers
Water rollers
Ink tray

Offset lithography Plate cylinder
Blanket cylinder
Printed letter

Damp area
Impression cylinder

▲ The gravure method is to etch away the area which is to be printed and to fill the etched area with ink. In this technique all text and illustrations have to be screened into dots.

▲ In lithography, water and ink are both applied to the printing plate. The surface properties of the plate are changed so the water sticks to some areas and ink to others.

165

Photography

- The finest microfiche is able to display up to 6,000 frames on a single sheet 10 cm by 15 cm. The whole telephone directory for a large city can be stored on one sheet.

- The fastest camera in the world is used in laser research at the Blackett Laboratory, London. It is able to take up to 33 billion images per second.

- The oldest surviving colour photograph was taken in 1877.

- Twinkling stars which appear white to the naked eye can be seen in full colour by means of long-exposure astrophotography.

- The original daguerreotypes needed up to 20 minutes' exposure in strong sunlight, but the film in a standard modern pocket camera is 36,000 times as fast, with an exposure time of only 1/30 of a second.

▶ Athletes at speed. By careful choice of camera lens and technique, the photographer can capture the spirit as well as the substance of the subject being photographed.

Photography is the science and art of permanently recording pictures of the world. A photograph can remind you of someone you love or bring back memories of things that are past. Photographs can be beautiful, impressive, informative or funny. The complicated chemistry of photography has been mastered so completely that photographs can be taken, developed and printed in less than a minute. Cameras have advanced so far that they can now operate completely automatically. Photography also allows us to look at things which we cannot see with our eyes alone because they move too quickly, too slowly, or are too far away. Photography is a means of mass communication also open to individuals.

166

Drawing with light

It is very simple to see an image of a brightly-lit scene on a screen. The screen must be in a dark box with a small pinhole in the box, through which the light from the outside world can come to illuminate the screen. If the pinhole is the right size, there will be an image of the scene outside focused on the screen. This was known thousands of years ago. The device was called a camera obscura which means "dark room". Although producing the image is simple, it is much more difficult to invent a screen which will permanently record the picture. This is the process of photography, which literally means "drawing with light".

The first ever photograph was taken in the year 1826 by a French soldier called Joseph Nicéphore Niépce. He recorded the view from his window on a metal plate, which had been coated with a light-sensitive substance. He had to wait all day for the picture to appear on the metal plate because the sensitivity of the coating on the plate was so low. Also, he did not use a lens to concentrate the light on to the metal plate. Soon afterwards he went into partnership with another Frenchman, who was called Louis Daguerre.

After Niépce's death, Daguerre invented a much more sensitive method of recording pictures, using copper plates coated with silver. The plates were first exposed to iodine fumes, which made the surface light-sensitive. The picture was taken and then the plates were exposed to mercury vapour, which brought out the picture on the plate. The picture was fixed permanently by washing the plate in salty water.

Because the new coating was more sensitive to light, pictures could be taken more quickly. The reduced exposure times allowed photographs of people to be taken for the first time. The portraits were called daguerreotypes after their inventor and were very popular.

▲ This daguerreotype picture of a street in Paris was taken by Daguerre himself in 1833. It includes the first photographed human figure. Early photographs such as this one required very long exposure times. The man in the picture stood still to have his boots cleaned for long enough for his image to be recorded on the photographic plate.

▶ A new photographic technique was invented in 1851, which became more popular than the daguerreotype. It was called the wet-collodion process and involved coating a glass plate with a liquid and then exposing it to the light. Photographers had to carry chemicals with them and coat the plate inside the camera.

TENT PACKED FOR TRAVELLING

Negatives and positives

The first photographers were trying to find a way of recording a picture as a "positive" image. They wanted a process which would record the bright areas of the screen as white in the photograph and dark areas as black. (All early photographs were black and white.) This proved to be very difficult.

A breakthrough was made by an Englishman, William Fox Talbot. He found a relatively simple way of first producing a "negative" of the photograph, and then making a positive picture from the negative. His method of printing was very similar to that still used today, and allows many prints to be made from a single negative.

Modern photographic film consists of a thin layer, or emulsion, of a chemical called a silver halide, on a transparent plastic sheet. The chemical is a combination of atoms of silver, and atoms of one of the elements bromine, chlorine, iodine or fluorine. These chemicals are sensitive to light. Normally transparent, they become black if they are exposed to light, or illuminated, so the film naturally records a negative picture in which the brightest areas of the scene produce the darkest area on the film.

A positive print can be made from the negative by placing the negative on a new piece of film and illuminating the film through the negative. This is called a contact print. The darkest areas of the negative produce the lightest areas on the print. If a print is required which is larger than the negative, the negative is held away from the film. Light shining through the negative then produces a larger image of it on the film.

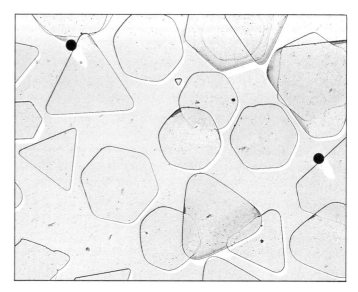

► (top) A highly-magnified view of a piece of unexposed photographic film, showing the crystals of silver halide. When light falls on the film, it produces some atoms of metallic silver within the crystals. (middle) Two crystals in which small areas have been blackened. If the exposed film was examined at this stage, there would be no picture to see. The film now has to be developed by immersing it in a chemical solution. In this solution the crystals blacken completely if they have been at all affected by light. (bottom) Crystals in the film after it has been developed. The picture is now visible on the film. The film is then immersed in a different solution which washes away all the unblackened crystals of silver halide. The film is no longer sensitive to light, and the picture is permanently fixed in the film.

168

Colour photography

Colour photography also uses crystals of silver halide as the light-sensitive material in which to record a picture. Colour film contains three layers of silver halide crystals. Each layer also contains chemicals which can form coloured dyes. When the film is exposed to light, each layer reacts only to red, green, or blue light. After the film has been developed, each layer contains coloured dye in the same places as the silver halide has been turned to silver. The dye colours are the complementary colours of those in the original image. For example, blue light results in yellow dye, and green light results in a magenta dye. The developed film therefore contains a colour negative of the image.

▼ In colour photography, the multi-layered film is developed using a chemical which produces different coloured dyes in the exposed parts of each layer. To print the image, white light is shone through the negative on to paper having a similar multi-layer coating. This reverses the colours, and the final print shows the same colours as the original image.

The colour negative process

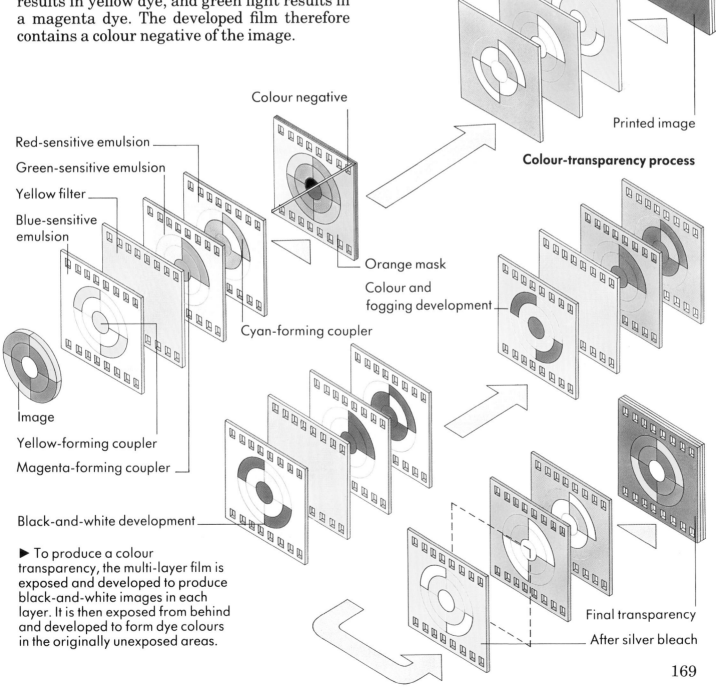

Printed image

Colour-transparency process

Colour negative

Red-sensitive emulsion

Green-sensitive emulsion

Yellow filter

Blue-sensitive emulsion

Orange mask

Colour and fogging development

Cyan-forming coupler

Image

Yellow-forming coupler

Magenta-forming coupler

Black-and-white development

Final transparency

After silver bleach

► To produce a colour transparency, the multi-layer film is exposed and developed to produce black-and-white images in each layer. It is then exposed from behind and developed to form dye colours in the originally unexposed areas.

169

Cameras

The first cameras were simple devices consisting of a box which held a single plate of film, and a lens to focus light on to the film. Modern cameras can be extremely sophisticated electronic machines with many functions automatically controlled. However, the principles of the camera are still the same. A roll of film loads into the back of the camera. Inside the camera it is completely dark until a picture is taken. A shutter inside the camera then opens to allow light coming through the lens to fall on the film. After a short time the shutter closes and the film can then be wound on to move an unexposed piece of film into position behind the shutter and lens. The main controls available to the photographer are the lens position, the lens aperture size and the exposure time.

Light from the scene to be photographed is focused on to the film by the lens, or group of lenses, to produce a sharp image of the subject.

However, light coming from objects at different distances will focus at different positions behind the lens. To focus on subjects closer to the camera, the lens body is rotated, which screws it away from the camera body and the film. The aperture size is set by rotating a ring at the back of the lens.

Reducing the aperture will reduce the amount of light reaching the film. To compensate, the exposure time (the time for which the shutter is open) can be varied between a few seconds, for very dark scenes, to one-thousandth of a second. The very fast speed is used to photograph fast-moving objects so they do not appear blurred. Most cameras can automatically set the exposure time after the aperture size has been chosen. The amount of light in the scene is measured inside the camera, and the exposure time is set to allow the same amount of light always to hit the film.

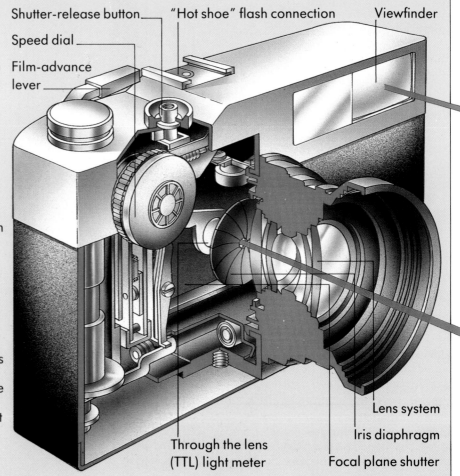

Pocket camera

The modern pocket camera is designed to be easy to use. The photographer composes a picture in the viewfinder and presses the shutter-release button. The shutter opens for a fraction of a second to expose one frame of the film, which captures an image of the scene viewed. Working the film-advance lever moves another frame of the film into position behind the shutter. The time of exposure is calculated automatically. A built-in light meter records the light level, and, acting on this information, the camera alters the shutter speed and/or lens aperture to give the correct exposure. The lens aperture is set by the iris diaphragm. This is made up of overlapping metal leaves, which advance or retract, depending on whether less or more light needs to enter. In low-level lighting conditions flash must be used to give adequate exposure. A flash gun is fixed into the "hot-shoe", which has electrical contacts that cause the flash to be set off when the shutter opens.

Shutter-release button
Speed dial
Film-advance lever
"Hot shoe" flash connection
Viewfinder
Through the lens (TTL) light meter
Lens system
Iris diaphragm
Focal plane shutter

► This is an early camera, built in 1856, which used the collodion process. A plate of glass would be fitted inside the back of the camera to record a single picture. The lens could be moved to focus the image. The chemicals needed to sensitize the glass were then put in the top of the camera. After the plate had been exposed to the image, further chemicals would be squeezed into the camera to process the photograph.

▼ The main type of camera in use today is the single lens reflex (SLR) camera. The SLR camera has only one lens (although this may consist of many optical components). The SLR camera can be fitted with many different types of lens: zoom, telephoto, macro, fisheye, which allow great variation in the focusing distance and in the angle of view of the camera. Many designs have automatic control of aperture size or shutter speed.

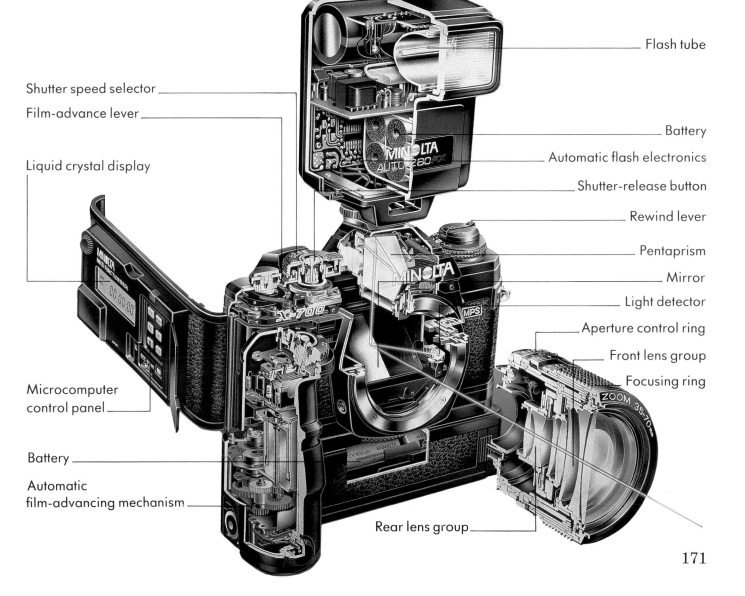

Flash tube

Battery

Automatic flash electronics

Shutter-release button

Rewind lever

Pentaprism

Mirror

Light detector

Aperture control ring

Front lens group

Focusing ring

Shutter speed selector

Film-advance lever

Liquid crystal display

Microcomputer control panel

Battery

Automatic film-advancing mechanism

Rear lens group

Advanced techniques

Photography was invented as a way of permanently recording scenes that people could see. Modern photography allows us to do much more than this. Pictures can be taken of things that we are unable to see. The normal human eye is sensitive to only a small range of the wavelengths of light. The shortest wavelength we can see is violet light; the longest is red light. Photographic film is not limited to this range. It can be designed to be sensitive to infrared wavelengths (longer than red) and ultraviolet wavelengths (shorter than violet). Some film can record wavelengths even shorter than ultraviolet, of which the most common to be used in photography are called X-rays.

Photography also allows us to overcome other limiting features of the human eye. Some things happen too quickly to be seen, other things happen too slowly. High-speed photography and time-lapse photography make these events visible. High-speed photography involves taking a large number of pictures in a short time. It is important that in each picture, the fast-moving object appears to be stationary so that the image is not blurred. This is achieved by using a flashing light as the illumination for the object. Each flash is a very short burst of bright light, which can repeat thousands of times a second.

Time-lapse photography involves taking frames of a film with long intervals between them. The film sequence is then played back at a much faster rate to speed up the apparent motion. This is a popular technique for showing such things as the growth of plants.

The human eye is also unable to see in very low levels of light. However, using photography, it is possible to build up a picture on sensitive film by collecting light over many hours. All the light hitting the film is recorded and contributes to the finished picture. This method is used in astronomical photography, for example in photographing dim and distant objects in the sky.

▼ A dragonfly in flight captured by high-speed photography. The exposure time is so short that the rapidly moving wings appear motionless.

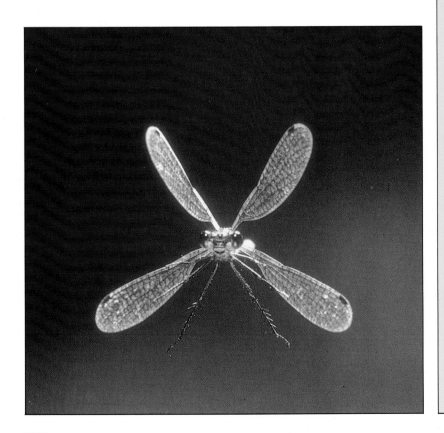

X-rays

X-rays are used to photograph internal regions of the body. X-rays are a form of electromagnetic radiation with a wavelength roughly one-thousandth that of visible light. The X-rays pass easily through soft tissue but are partly blocked by teeth and bones. Images of the teeth and bones show on X-ray film placed behind a body.

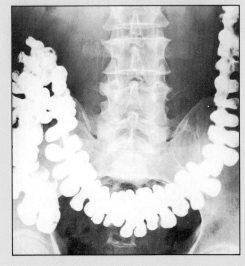

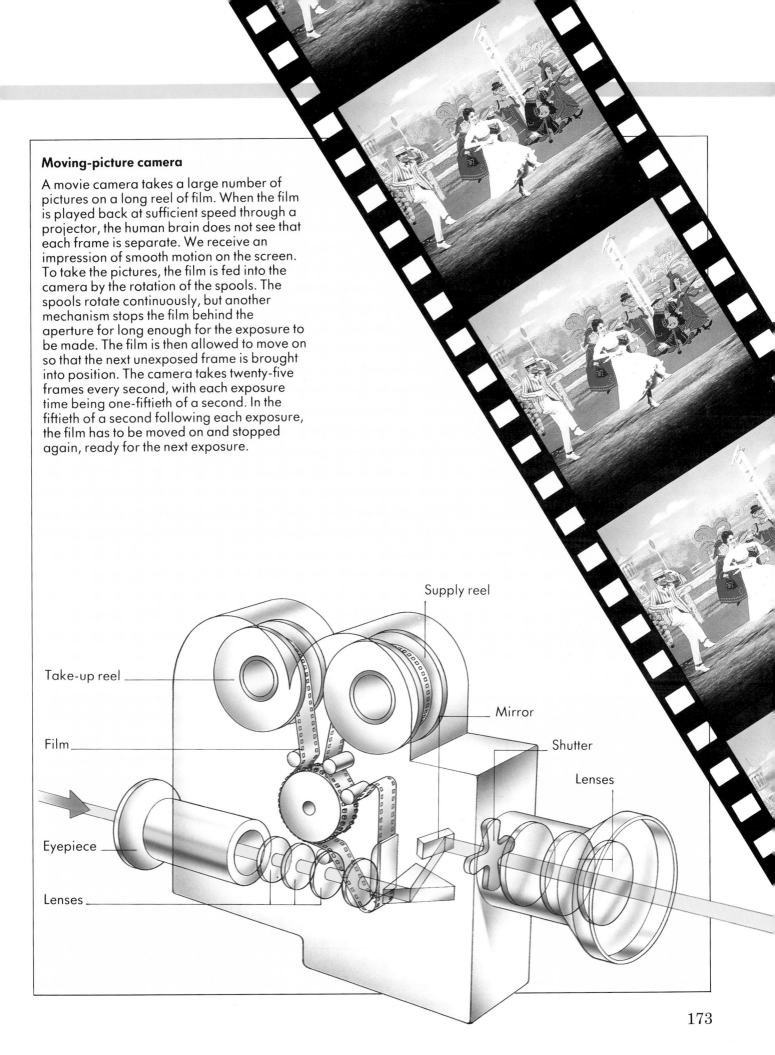

Moving-picture camera

A movie camera takes a large number of pictures on a long reel of film. When the film is played back at sufficient speed through a projector, the human brain does not see that each frame is separate. We receive an impression of smooth motion on the screen. To take the pictures, the film is fed into the camera by the rotation of the spools. The spools rotate continuously, but another mechanism stops the film behind the aperture for long enough for the exposure to be made. The film is then allowed to move on so that the next unexposed frame is brought into position. The camera takes twenty-five frames every second, with each exposure time being one-fiftieth of a second. In the fiftieth of a second following each exposure, the film has to be moved on and stopped again, ready for the next exposure.

Take-up reel

Film

Eyepiece

Lenses

Supply reel

Mirror

Shutter

Lenses

173

Telecommunications

Spot facts

- Transmission along the latest optical fibres can accommodate up to 20,000 telephone calls at a time.

- The craze for "videotex", or screen-based information systems, has been very successful in France. Over 10 million households have a miniature video terminal, or Minitel.

- Early Morse sets were able to transmit about 35 words per minute.

- The largest ever private telegraph system was operated by the US armed forces, with 2,700 centres in North and South America and 1,600 other centres around the world.

From the telephone in any house or office in a major city, a caller can dial a series of numbers and in a few seconds be connected to any telephone in one of a hundred different countries of the world. The telephone network provides a means of communication which is personal. Printing, television, radio and recording are means of mass communication. The telephone system works through a vast network of cables which cross countries, continents and oceans. Microwave stations beam our telephone calls through the air. Satellites are used to relay calls at the speed of light from one country to another. Telephone lines can also be used to carry digital information between computers.

▶ Talking on the telephone is the most personal form of telecommunication. Telephones can bring together in an instant people who may be thousands of kilometres apart at the opposite ends of the Earth.

174

Signals by wire

Electricity was first used in machines in the first half of the last century. One of the first useful applications of electricity was the construction of a device which could make the voltage at the end of a long wire the same as the voltage at the beginning. Normally the voltage would decrease along the wire because of the electrical resistance of the wire. The new system worked by using the low-voltage strength remaining at the end of the cable to trigger an amplifying device. This boosted the signal back to its original strength to start the journey towards the next amplifying station. A scheme like this is necessary for all long-distance communications whether they are by wire and optical fibre.

The first electrical communications machine used two magnetic needles (like the needle of a compass) which were deflected in different directions, depending on the voltage signals it received along the wires from the sending machine. The same machine could both send and receive signals. The operator had to learn that each letter of the message was represented by a certain number of needle deflections.

The next advance was made by Alfred Vail and Samuel Morse, who installed the first of their communications systems in 1844. This used a code for each letter of the alphabet, consisting of a certain combination of dots and dashes. Common letters such as E and T were given short codes. Uncommon letters were given the long codes. (For example T is "–", while Q is "–.–".) The operator sending the message would translate it into this Morse code during transmission and would tap out the code directly on an electrical contact of the sending device. The receiving device made a click for each tap of the contact. Two clicks close together indicated a dot, two further apart indicated a dash. Very high speeds could be achieved by skilled operators.

Many businesses have telex machines to provide a written record of telecommunications. A telex machine consists of a keyboard, a display screen and a printer. The operator types the message and can check it on the screen. The message is transmitted as a series of on-off voltage pulses. The receiving machine decodes these pulses and prints out the message.

▼ An early design of Morse code receiver. The electrical signals arriving at the receiver would make a pen write dots or dashes on to paper. Later designs simply made clicks for dots and dashes and the receiving operator wrote the message out.

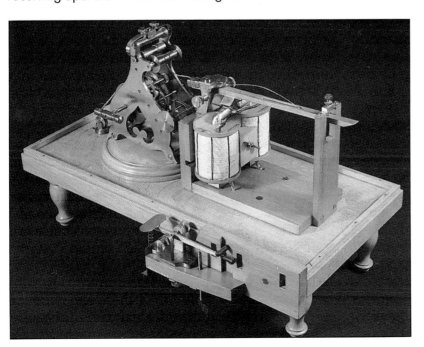

▲ A two-needle telegraph, invented in England in 1837. This was the first practically instantaneous communications method between people who could not see each other. Experienced operators could read up to 20 words a minute.

The telephone

When you speak, you make sound waves in the air. The sound waves can be detected by the human ear so other people can hear what you are saying. The telephone is an instrument which can also respond to the sound waves of your voice and convert them into electrical signals. The signals travel along wires to the receiving telephone, where they are converted back into sound waves. These sound waves have to be similar to the original ones so that the person listening can understand what is being said.

The handset of a telephone consists of a microphone at one end and a loudspeaker at the other. When you speak into the microphone, the impact of the sound waves makes the surface vibrate. Behind the surface of the mouthpiece is a small chamber full of pieces of carbon. As the surface vibrates, the volume of the chamber changes and the carbon granules become more or less tightly packed. This changes the elect-

rical resistance of the carbon. A current flows through one of the wires that comes into the telephone. As the carbon resistance changes, the electric current in this wire also changes. It is this electric current which carries the record of the voice of the person using the phone. The changing electric current travels down wires and is routed by the telephone exchange to the right telephone.

When the current arrives at the receiving phone, it passes through coils wrapped round a magnet in the earpiece. The changing current produces in turn a changing magnetic field in the receiver. This either pulls or pushes a diaphragm, positioned behind the earpiece of the receiver. The vibrating surface reproduces the sound waves which were originally responsible for the varying current. Although the reproduction is not perfect, somebody listening can usually understand what was said into the calling telephone.

▲ Early telephone exchanges were run by operators connecting calls by hand. This was a slow process, and prone to mistakes.

▶ Modern telephone exchanges are all electronic. Connections to the dialled telephone number are made automatically. Human intervention is only needed for maintenance.

Telephone network

▶ At the receiving end of the telephone, a changing magnetic field vibrates a diaphragm in response to the electric signals it receives. The diaphragm sets up sound waves in the air which closely resemble those spoken by the caller.

▶ Every telephone in the world has a number. Each telephone is connected by wires to a local exchange. In a large city the local exchanges are all connected to a central exchange for the city. Each city or area can be called from every other area.

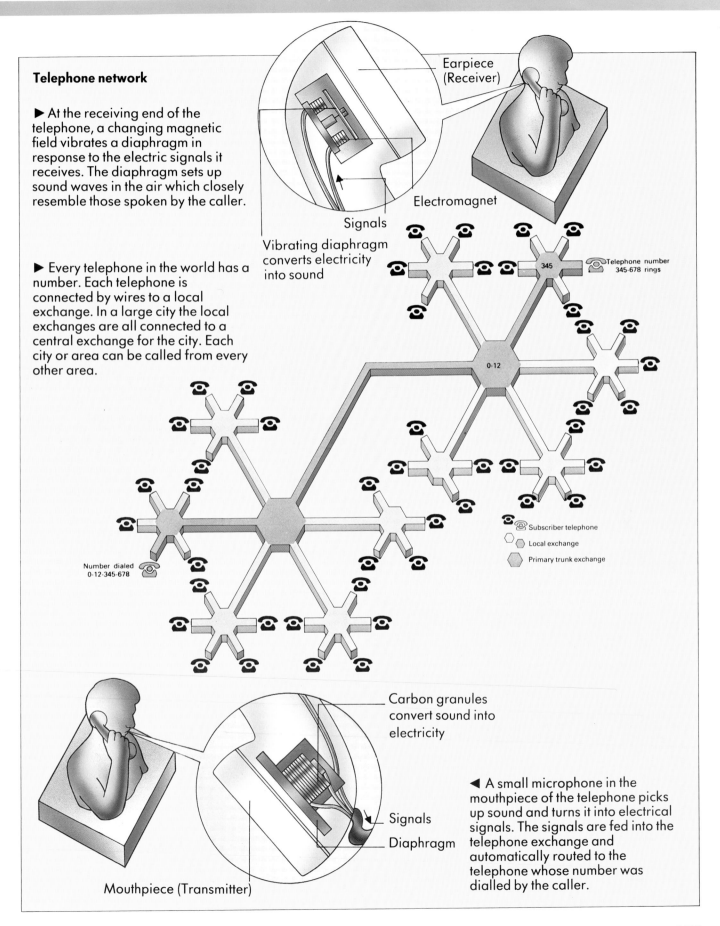

Earpiece (Receiver)

Electromagnet

Signals

Vibrating diaphragm converts electricity into sound

Telephone number 345-678 rings

Subscriber telephone

Local exchange

Primary trunk exchange

Number dialed 0-12-345-678

Carbon granules convert sound into electricity

Signals

Diaphragm

Mouthpiece (Transmitter)

◀ A small microphone in the mouthpiece of the telephone picks up sound and turns it into electrical signals. The signals are fed into the telephone exchange and automatically routed to the telephone whose number was dialled by the caller.

Cable links

Individual telephones are connected by a two-wire cable to a distribution point, usually on a pole in the street. These distribution points are in turn connected to a local cabinet. This combines all local lines into main cables running under the road to the exchange. Local exchanges throughout the country are connected by coaxial cable. This consists of a central conductor surrounded by an electrically insulating plastic which is itself surrounded by a conducting cylinder.

Multiplexing

Because of the expense of laying cables and the increasing demand for telephone lines, it has been important to develop ways of using a single cable for many telephone calls at a time. This is called multiplexing. A common system of multiplexing involves pulse code modulation (PCM). Instead of sending the complete telephone signal along the cable, the strength of

the signal is measured eight thousand times a second. This is called digitizing the signal. Each value of signal strength is represented as an eight-digit binary number consisting of ones and zeros. These numbers are sent along the cable and the original signal is reconstructed at the receiver. Because the binary number can be transmitted very rapidly, the cable is only being used for a very short period every one eight-thousandth of a second to transmit information for one telephone call. It is possible to send many other calls at the same time along the same cable by interleaving signal strength

▼ A chaotic mass of telephone cables in 1909 Kansas, United States. When telephones were first invented, every telephone had to be connected to every other telephone by two wires. Even after the installation of local exchanges, every simultaneous call into the exchange had to arrive by a different wire. The solution to this problem arrived with the invention of multiplexing.

Fax and electronic mail

Both facsimile (known as fax) and electronic mail (e-mail) are ways of sending documents without using the conventional postage system. Both are much faster than ordinary mail, and both make use of telephone lines to transmit information. Fax and e-mail use a device called a modem (short for modulator-demodulator). The modem converts digital information from a fax machine or a computer into a sound signal which can be sent down a telephone line. A modem at the other end of the line changes this signal back to the original format.

The fax machine can transmit any document, whether it contains words or pictures. The document is placed face down on the machine which then pulls it over a strip of bright light. A scanner measures the reflectivity of the paper. White areas reflect the light much more than areas with writing or drawing on them. This information is recorded in a sequence of numbers. For each small area of paper, a one or a zero is added to the sequence depending on whether the page is light or dark at that point. This information is then sent along a telephone line to the receiving fax machine, where it is used to copy the original page.

E-mail provides a way for people with computers to communicate with each other. A computer can be linked to the telephone system using a modem. The sender types out the message and instructs the computer to send it. The information is converted into a signal for transmission down a telephone line. At the destination computer, the signal may be reconverted and shown on the receiver's screen at once. Or if that computer is turned off or in use, the message can be stored in the central memory bank for later retrieval. By this method messages can be transmitted purely as electronic signals, without ever being printed.

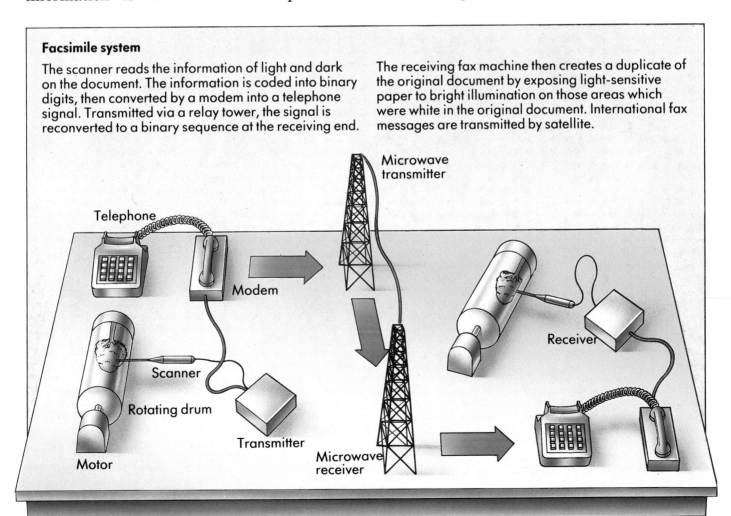

Facsimile system

The scanner reads the information of light and dark on the document. The information is coded into binary digits, then converted by a modem into a telephone signal. Transmitted via a relay tower, the signal is reconverted to a binary sequence at the receiving end.

The receiving fax machine then creates a duplicate of the original document by exposing light-sensitive paper to bright illumination on those areas which were white in the original document. International fax messages are transmitted by satellite.

Microwave transmitter

Telephone

Modem

Scanner

Rotating drum

Transmitter

Motor

Microwave receiver

Receiver

On the air

Spot facts

- *The first use of the radiomicrophone was in 1947. Reg Moores, an entertainer, used it in an ice-pantomime in Brighton, England.*

- *Up to 30 images, or 6,000 billion billion electrons, are normally transmitted by television every second. Images persist in the human eye for one-tenth of a second, so television must produce more than ten images per second to give the impression of a moving image.*

- *In the United States, any one television station uses as much space on the airwaves as all the AM radio channels put together.*

- *The first videophones appeared in the early 1970s. As the speaker's voice was relayed by telephone, his or her image was transmitted on a screen by closed-circuit television.*

▶ The scene in a television broadcasting studio during the screening of the news and weather. A number of cameras are used to view the newscasters, who read from notes and from teleprompt screens.

Television and radio are the most used forms of mass communication. They are also the main source of entertainment in the home. Television and radio are broadcast from transmitters, which send their signals through the air in all directions. The signals can be picked up by an aerial attached to an individual set.

The advent of television has changed the view we have of the world, making it seem smaller. Live television broadcasts allow us to see events as they happen. We can watch momentous world events from the other side of the planet. We can have the best view in the sports stadium, and hear commentary on the game at the same time. Satellites which orbit the Earth at a height of over 35,000 km are used to relay television signals from distant places into our homes.

Wireless telegraph

About one hundred years ago it was discovered that electric and magnetic energy could travel through the air, or even through empty space. This is called electromagnetic radiation. It was calculated that the speed at which this radiation should move was the same as the speed of light. It was realized that visible light is a kind of electromagnetic radiation, but that there are very many other kinds of electromagnetic radiation which are invisible to humans. These include what we now call gamma rays, X-rays, ultraviolet and infrared radiation, microwaves, and radio waves.

In all early forms of electrical communication, information was carried by an electric current travelling along a wire. Following the discovery of electromagnetic radiation, some people saw the possibility of using it to create a form of communication without wires. This wireless telegraph was developed in particular by an Italian called Guglielmo Marconi. His first working system used electric sparks to generate the electromagnetic radiation, and a coil of wire which could be attached to an earpiece to detect the radiation. The radio waves radiated from the spark in all directions. Some passed through the coil which generated a small current in the loop of wire. This was used to create a faint clicking sound in the earpiece.

Using this system, Marconi could send messages in Morse code between people who were not connected by a wire. In 1896 he sent Morse code messages for a short distance between a spark generator and a receiver. In 1901 he sent messages for nearly 5,000 km across the Atlantic Ocean. It surprised everybody that the radio waves could be received from so far away. It was later found that radio waves are reflected by the ionosphere and reach the Earth many thousands of kilometres away.

Electromagnetic waves

Electric field

Magnetic field

Direction of movement

▲ Radio waves, X-rays, and visible light are all forms of electromagnetic radiation. The radiation consists of oscillating electric and magnetic fields which are at right-angles to each other. The waves can move through empty space. Electromagnetic radiation moves at just less than 300,000 km/sec.

◄ Guglielmo Marconi, who invented the first successful wireless telegraph. In 1901 he transmitted the Morse code signal for the letter S from Cornwall, England to Newfoundland, Canada across the Atlantic. He was awarded a Nobel prize in 1909.

183

Radio

Radio waves are a type of electromagnetic radiation. All electromagnetic radiation has a wavelength and a frequency. The wavelength is the distance between the peaks of the wave. The frequency is the number of times the wave goes up and down in a second. Wavelength is measured in metres; frequency is measured in hertz (Hz), kilohertz (kHz) or megahertz (mHz). A megahertz is a thousand kilohertz.

The first transatlantic radio messages of Marconi consisted of radiation at all frequencies which were sent in three short bursts to spell out the letter S in Morse code. With improvements in electronics at the beginning of the century, it became possible to produce radio waves at almost a single frequency. Radio receivers were also invented which could be tuned to detect a signal at a single frequency.

These two developments allow many radio stations to broadcast each on a different frequency for the listener to choose from. Radio broadcasts are made in three ranges of frequency. Long-wave signals have a frequency between 150 and 280 kHz. Medium waves lie between 540 and 1,600 kHz. The highest band, called FM, is between 88 and 108 MHz.

A radio system consists of a microphone linked to a transmitter, and a receiver linked to a loudspeaker. When someone speaks or plays music into the microphone, the sound waves are converted into electrical signals. These signals are then mixed with a radio wave. This is called modulation. The radio wave is called the carrier wave.

The radio wave with all the sound information on it is then beamed in all directions from high-power transmitters. A radio aerial can detect these radio waves passing across it. The aerial detects all frequencies but by tuning the radio to a particular position only one frequency is amplified within the radio. The radio subtracts the carrier wave from the signal which leaves the original sound information. This is used to drive a loudspeaker, which makes the sound from the radio audible.

▼ Radio signals controlling model aircraft. The joystick controls produce different signals depending on the direction of movement. The large aerial sends out the signal which is picked up by a receiver on the plane. This controls the movement of wings and rudder.

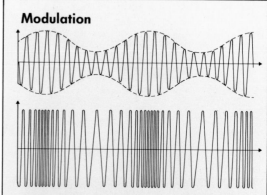

Modulation

Modulation is the process of adding information on to a radio wave. A pure radio wave has a constant frequency and a constant signal strength (amplitude). It is possible either to change the amplitude of the carrier wave or to change its frequency to make it carry sound information to a radio receiver. These two methods are called amplitude modulation (AM) and frequency modulation (FM).

Cellular radio

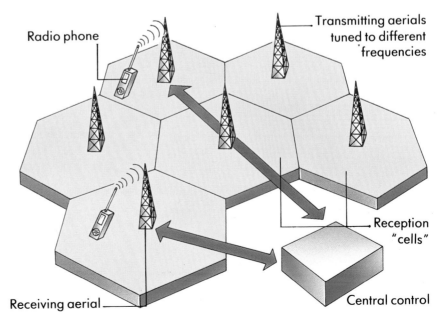

▲ A local-radio DJ (disc jockey) speaks to a caller on a popular "phone-in" programme. Local-radio stations broadcast from FM transmitters, which have only a limited, above-the-horizon range.

Radio phone

Transmitting aerials tuned to different frequencies

Reception "cells"

Central control

Receiving aerial

▲ Using a portable cellular telephone, a cross between a telephone handset and a "walkie-talkie" radio.

▶ In a cellular radio system, telephones link into the telephone network via local transmitters.

185

Television

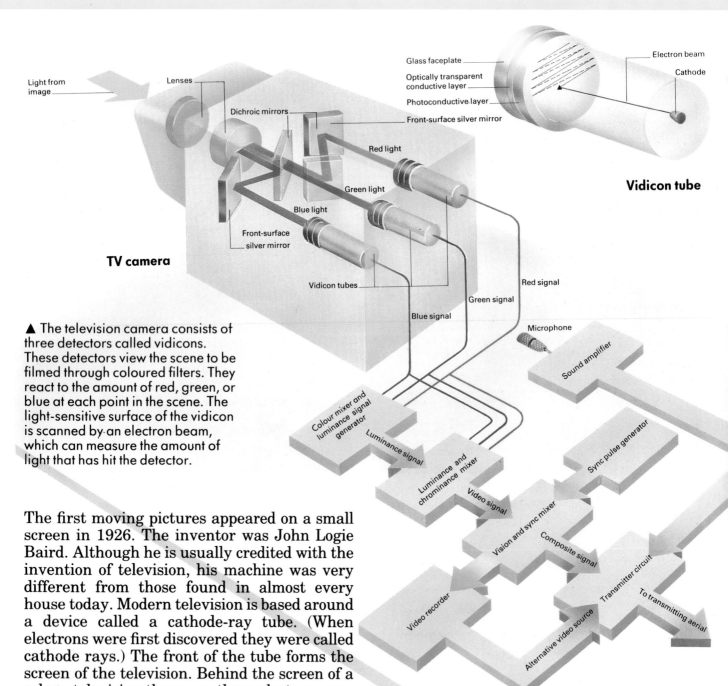

Vidicon tube

Labels: Glass faceplate, Optically transparent conductive layer, Photoconductive layer, Front-surface silver mirror, Electron beam, Cathode

Labels (TV camera): Light from image, Lenses, Dichroic mirrors, Red light, Green light, Blue light, Front-surface silver mirror, Vidicon tubes, **TV camera**

Labels (signal flow): Red signal, Green signal, Blue signal, Microphone, Sound amplifier, Sync pulse generator, Colour mixer and luminance signal generator, Luminance signal, Luminance and chrominance mixer, Video signal, Vision and sync mixer, Composite signal, Transmitter circuit, To transmitting aerial, Video recorder, Alternative video source

▲ The television camera consists of three detectors called vidicons. These detectors view the scene to be filmed through coloured filters. They react to the amount of red, green, or blue at each point in the scene. The light-sensitive surface of the vidicon is scanned by an electron beam, which can measure the amount of light that has hit the detector.

The first moving pictures appeared on a small screen in 1926. The inventor was John Logie Baird. Although he is usually credited with the invention of television, his machine was very different from those found in almost every house today. Modern television is based around a device called a cathode-ray tube. (When electrons were first discovered they were called cathode rays.) The front of the tube forms the screen of the television. Behind the screen of a colour television there are three electron guns which use a high voltage to speed up electrons and accelerate them towards the screen.

When electrons hit a point on the screen they make that position glow with a colour which depends on a chemical with which the screen is coated. A colour television screen is coated with three different chemicals (called phosphors) which glow red, green or blue. The picture is built up from dots or lines of only these colours. From a normal viewing distance these primary colours merge to reproduce the colour in the scene being televised. From close to the screen,

▲ The signals from each of the vidicon light detectors in the television camera are called the chrominance signals because they measure the amount of each colour. The signals are also used to measure the overall brightness of each part of the scene. This is the luminance signal. The chrominance and luminance signals are combined with the sound signal before being transmitted from the aerial.

the individual areas of primary colour are easily visible.

The screen is divided into many hundreds of horizontal lines of phosphors: 625 lines in the UK and Europe, 525 in the US and Japan. The electron beams are scanned across the screen, illuminating only a single point on the screen at any instant. The electrons are deflected by magnets, which scan the beam from left to right across the first line. The beam then jumps back

across the screen and scans the third line, also from left to right. The scanning continues on every other line until the bottom of the screen is reached. This takes one-fiftieth of a second. The beam then jumps back to the second line and fills in the even-numbered lines in the same way. A complete screen is drawn in one twenty-fifth of a second. This scanning rate of twenty-five frames a second is enough for the scanning to be invisible to the human eye.

▼ The signal is picked up by the aerial. It is passed into the TV set where it is split into its constituent parts. The sound signal goes to the loudspeaker. The luminance and chrominance information goes to the electron guns, which illuminate the phosphors on the screen so as to produce a colour image of the scene as viewed by the camera.

▼ The three electron beams are fired towards the screen. The screen is coated with lines of blue, green and red phosphors which glow with colour when they are energized by the electrons.

TV receiver

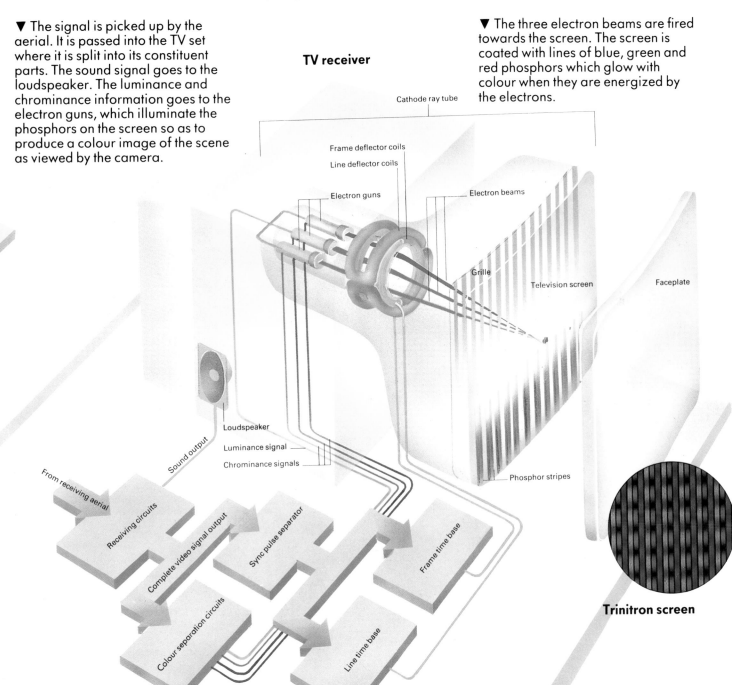

Cathode ray tube

Frame deflector coils

Line deflector coils

Electron guns

Electron beams

Grille

Television screen

Faceplate

Loudspeaker

Luminance signal

Chrominance signals

Phosphor stripes

Sound output

From receiving aerial

Receiving circuits

Complete video signal output

Sync pulse separator

Frame time base

Colour separation circuits

Line time base

Trinitron screen

On screen

Television is an extremely popular source of entertainment and information. In Western Europe there is a television in most homes. In the United States there are almost as many televisions as there are people in the country. The widespread ownership of television sets and the enthusiasm of people to watch them make broadcasting on television a very efficient way of reaching the attention of millions of viewers. Companies pay large amounts of money to advertise their products in breaks during and between programmes.

In most countries the broadcasting organizations are separate from the government, and many are independent companies. By transmitting their programmes on cable networks or via satellite, they provide many different channels.

A television aerial will pick up programmes sent through the air from powerful transmitting towers. Television programmes can also arrive in the home along cables or from a satellite. In a cable TV network, a wire runs from each house to a main cable. The main cable is connected to either a local receiving antenna or directly to the television station. Cable TV became widespread in cities in the United States, where the many tall steel buildings make conventional TV reception difficult.

Television signals can also be transmitted from the TV station to one of the communication satellites that orbit the Earth. The satellite receives the transmissions and sends them back to Earth, radiating the signal over a large area of land. The programmes are picked up by means of a small dish aerial.

In addition to broadcasting programmes, television can display pages of information. The information is sent by the TV company but the TV owner has to install an electronic unit which decodes the signal and displays the text. A similar system called Viewdata allows the TV user access to large banks of information via a telephone line. The databank can be questioned and the answer displayed on the screen.

Cable television

People who subscribe to a cable-TV network watch television programmes received, not via an external aerial, but via cable. The network control station (picture right) receives programmes beamed down to it from one or more communications satellites.

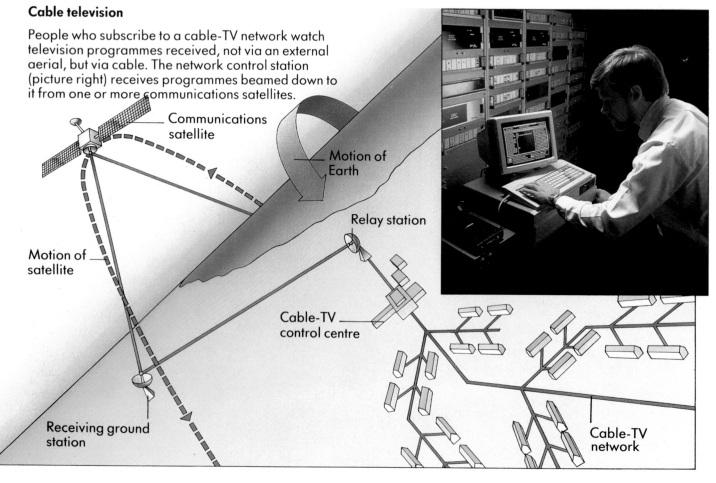

Communications satellite

Motion of Earth

Motion of satellite

Relay station

Cable-TV control centre

Receiving ground station

Cable-TV network

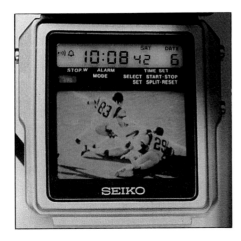

▲ A watch-TV, which uses liquid crystal display (LCD) to create a black-and-white picture. The screen is made up of liquid-crystal cells, which change their reflectivity when an electric field is applied to them. Other new technologies are leading to the development of full-size flat-screen TV receivers.

▼ Closed-circuit television (CCTV) being used at the Live Aid pop concert in London in 1985 to give the audience a better view of the performers. The concert was also relayed worldwide by satellite.

▲ The weather "page" of the Ceefax information "magazine" broadcast on television by the British Broadcasting Corporation (BBC). The BBC pioneered the system, called teletext, in the 1970s.

Recording

• *The nursery rhyme "Mary had a little lamb" was the first sound recording played in public in 1877. The reciter was the US inventor Thomas A. Edison, who recorded and played it back on his "favorite invention", the phonograph.*

• *Early phonographs used cactus needles for styli. The tip radius of the stylus of old 78 rpm record players was 0.007 cm. The modern playback stylus has a tip radius of 0.0005 cm.*

• *Modern manufacturing techniques can produce long-playing discs with a surface accuracy of more than 0.00025 cm.*

• *A compact disc can still reproduce music perfectly, even when it has a hole drilled through it.*

▶ Young people gyrate to the throbbing music at a discotheque in Disneyland, Los Angeles, USA. Disco dancing to records remains as popular as ever. But for personal listening, cassette tapes are now preferred.

Sound can be recorded and replayed in many different ways. The methods which are in everyday use are magnetic tape, record and compact disc. To record on tape, the sound is converted to electric signals, which are used to alter the magnetism of the tape. To manufacture a record, the sound vibrations are fed to a cutter, which makes a wavy groove in a plastic disc. For the compact disc the sound is converted to a string of numbers, which represent the sound waves. The numbers are recorded on the disc and "read" by laser.

The more difficult problem of recording picture information can also be solved using either magnetic tape (video) or compact disc. The main use of these inventions is in home entertainment and in broadcasting.

Microphones and speakers

Sound waves are a variation in air pressure. Microphones have to use this varying air pressure to produce a similar variation in a voltage or current within the microphone. In that way, electrical signals can be sent along wires from the microphone. There are two main types of microphone in use today. One is the dynamic microphone, the other is the condenser microphone. Both types have a thin sheet, or diaphragm, behind a protective front cover.

In the dynamic microphone, there is a coil of wire attached behind the diaphragm. A movement of the diaphragm in response to the varying air pressure of sound waves also moves the coil past a magnet in the microphone. This creates a varying electric current in the coil. In the condenser microphone, the diaphragm movement changes the spacing between it and a second sheet positioned behind it. This changes an electrical property called capacitance. As the capacitance changes, electric current travels along the wires from the microphone.

▼ Microphones convert the pressure variations of the sound waves into similar variations in electric current. The movement of the diaphragm changes the distance between the diaphragm and a charged plate. This changes the amount of charge the plate can hold as provided by the battery. As it changes, current is moved along the wires from the microphone.

Microphone

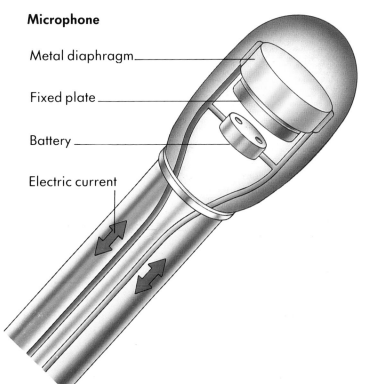

Metal diaphragm

Fixed plate

Battery

Electric current

Loudspeakers

Loudspeakers have to convert a varying electrical signal into sound. They do this by using the signal to move of the thin cone which forms the front of the speaker. The movement sets up air pressure variations, which reproduce the original sound. The most common design of speaker operates in a similar way to the dynamic microphone. A coil of wire, which is attached to the speaker cone, sits in the field of a magnet. The electric signal is fed into the coil which causes it to move in the magnetic field. This makes the speaker cone vibrate to match the oscillations of the electric current.

To reproduce accurately the original sounds, it is important to have a range of speakers of different physical size. Small speakers (called tweeters) are good for reproducing high-pitched sounds. Large speakers (called woofers) are best for low-pitched sounds. The accuracy of sound reproduction is called fidelity. Very high fidelity (hi-fi) speakers have three cones, each for a different frequency range.

Loudspeaker

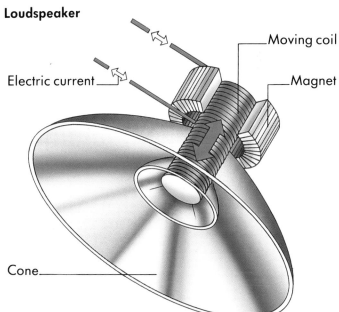

Moving coil

Magnet

Electric current

Cone

▲ When the varying electric current reaches the loudspeaker, it is fed to a coil. The coil sits in a magnetic field. Changes in the current alter the amount of force which the magnetic field exerts on the coil. The coil moves backwards and forwards inside the magnet. The coil is attached to a thin paper cone. The cone moves the air around it to reproduce the pressure variations in the original sounds.

Disc and tape

Sound can be permanently recorded and played back using either magnetic tape (reel-to-reel and cassette) or plastic discs (records). To record on a magnetic tape, the electric signal from a microphone is fed to an electromagnet which forms the main part of the recording head of a tape recorder. The tape is drawn past the recording head as it is wound from one spool to the other. It is very important that this movement is smooth. When playing back the tape, it has to move at the same speed as that of the recording. The tape has a thin layer of tiny metallic crystals of iron oxide. Each one of the crystals can act as a magnet. When the tape passes over the recording head, the crystals are magnetized so that the poles point in a particular direction. When the tape is played back the process is reversed. The magnetization of the crystals induces currents in the replay head, which are passed to a loudspeaker to reproduce the original sounds.

Recording on magnetic tape is also the first step in making a recording on disc. The taped sounds are fed to a cutter which scores a groove in the disc. The waviness of the groove contains the information about the volume and pitch of the original sounds. This first disc is then used to produce a metal copy. The metal copy has ridges which exactly match the grooves in the

▼ A recent very successful development in tape technology is the personal stereo, often called a walkman. Great effort was invested by many Japanese companies, particularly Sony, in the miniaturization of the tape player and in the design and appearance of the product.

The pioneers

Towards the end of the 1800s two systems of sound reproduction were invented. The reel-to-reel system, invented by Valdemar Poulsen, stored sound on a magnetized steel wire. The wax cylinder system was invented by Thomas Edison (right). It stored sound in grooves in the wax.

original plastic (acetate) disc. Two metal copies are required, one for each side of the record. To produce a record, blank plastic discs are pressed between the two metal masters. This produces a groove in each side of the disc, which is a copy of the original. When the stylus of the record player runs in the groove, its vibrations are picked up and translated into electrical signals. The signals are amplified and fed to the speakers of the system.

To make a stereo recording on a disc, the groove is produced by two cutters. Each cutter forms the shape of one side of the valley of the groove. When the disc is played, the effect on the stylus from the two differently-shaped sides of the groove can be split up in the stylus cartridge. Two signals are sent from the pick-up arm to the amplifier. The amplified signals are then sent to separate loudspeakers, which together reproduce stereophonic sounds.

Record

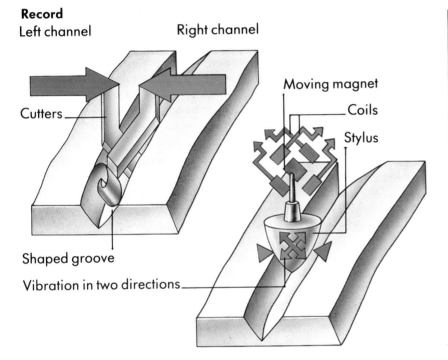

Left channel

Right channel

Cutters

Shaped groove

Vibration in two directions

Moving magnet

Coils

Stylus

◀▲ Sound is stored on plastic discs in the shape of the groove which is cut in the disc. The original sound is fed to cutters, which shape a long spiral groove in a disc of acetate. Stereo sound is reproduced by cutting the two sides of the groove with different shapes. The sound is reproduced by a diamond stylus (photo above) which runs in the groove and vibrates as it is moved by the wavy edges. Sound is stored on tape in the magnetism of a thin layer of iron oxide crystals which coat the plastic tape. Varying currents from a microphone change the strength of the magnetic field of the recording head. This magnetizes the crystals in particular directions. Running the tape past the replay head induces electric currents, which are fed to a loudspeaker. Stereo recordings can also be made on magnetic tape. The recording is made in two tracks at different positions across the tape. The sounds on each track are played back through loudspeakers.

Magnetic tape

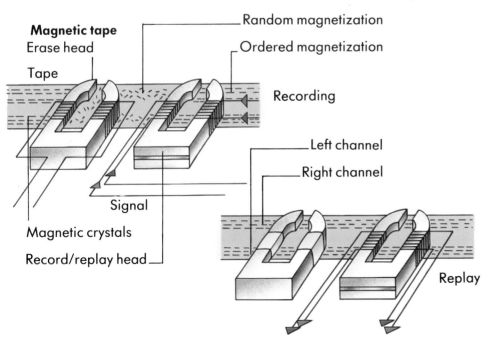

Erase head

Tape

Random magnetization

Ordered magnetization

Recording

Signal

Left channel

Right channel

Magnetic crystals

Record/replay head

Replay

193

Video

Television pictures can be stored on videotape in the same way that sound can be recorded on audio tape. The two kinds of tape are made from similar material: magnetized iron oxide particles on a plastic tape. But vastly more information has to be put on the video tape than on the audio cassette. Twenty-five times every second the video recording head has to put information on to the tape. This information is on the brightness and colour at each point of the 625 lines of the television screen.

The first video machines which used magnetic tape tried to use exactly the same recording technique as an audio tape recorder. The recording head was kept stationary and the tape was moved past it. Because of the enormous amount of information that had to be recorded, the tape had to be moved extremely quickly over the head. Speeds of up to ten metres per second were used. Modern video recorders move the tape at only two centimetres per second, five hundred times slower. The decrease in speed has been made possible by very great improvements in the way that television picture information is put on to the tape by the recording head.

Modern video recorders

Instead of moving the tape very quickly past the head, modern video recorders move the head very quickly past the tape by fixing it to a cylinder which is rotated at high speed. When a videotape is put into a video cassette recorder (VCR), the cover that protects the tape is lifted and two plastic posts rise behind the tape and pull it out of the cassette. The tape is wrapped around the recording head cylinder. The cylinder contains two recording/play heads which rotate with it. The cylinder is twice as wide as the tape and set at an angle to it. As the cylinder rotates beneath the angled tape, the recording heads write the information on the tape in long diagonal lines, which run from one edge of the tape to the other.

There are two main VCR systems: Betamax and VHS (Video Home System). Both use 19-mm wide tape.

▼ The use of a video camera recorder (camcorder) for recording family events and holidays is becoming more and more popular. Simple camcorders are available now which record picture and sound at the same time. Tapes recorded like this can be played back on standard VCR machines.

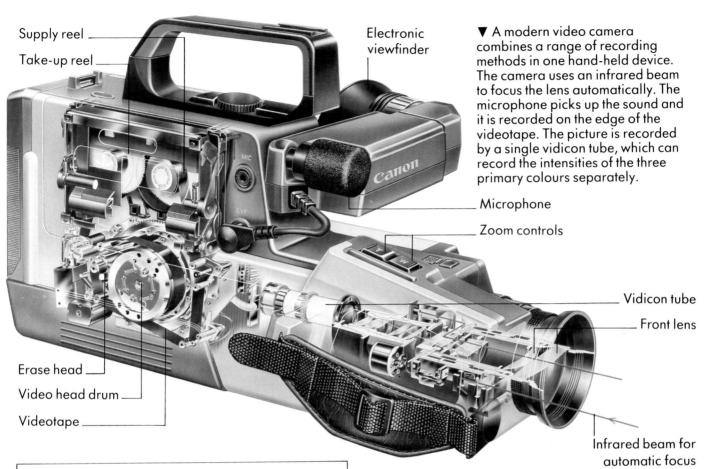

Supply reel

Take-up reel

Electronic viewfinder

▼ A modern video camera combines a range of recording methods in one hand-held device. The camera uses an infrared beam to focus the lens automatically. The microphone picks up the sound and it is recorded on the edge of the videotape. The picture is recorded by a single vidicon tube, which can record the intensities of the three primary colours separately.

MIC

EVF

Canon

Microphone

Zoom controls

Vidicon tube

Front lens

Erase head

Video head drum

Videotape

Infrared beam for automatic focus

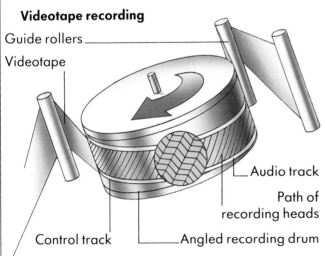

Videotape recording

Guide rollers

Videotape

Audio track

Path of recording heads

Control track

Angled recording drum

Two recording heads are positioned on the rotating cylinder. As they pass the tape they write diagonally across it because of the angle of the cylinder. Because of the different angles at which the recording heads are set, each head cannot read information from the tracks on either side. This reduces interference and allows more room for information on the tape.

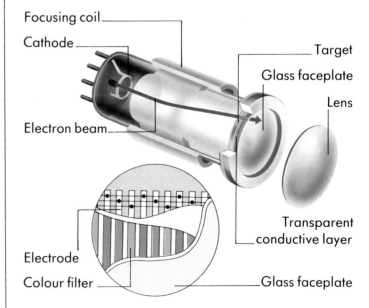

Focusing coil

Cathode

Target

Glass faceplate

Lens

Electron beam

Electrode

Colour filter

Transparent conductive layer

Glass faceplate

▲ This is a single vidicon tube which can measure the brightness of each of the colours red, green and blue. Behind the glass plate at the front, there is an array of very small filters which allow through only one of the colours. Behind the filter the electron beam can measure separately the light intensity of each colour.

Digital discs

The newest form of recording to reach the home is the compact disc. The discs themselves and the disc players require very precise and controlled manufacturing methods. This makes them probably the most complicated and advanced machines that are in everyday use in the home.

There are three main features of compact discs which make the quality of the sound obtainable from them superior. The first is that sound is recorded digitally on them, and played back digitally. The second is that unlike the stylus of a record player, or the recording head of a tape recorder, the device used to "read" the sound does not actually touch the disc. This removes the danger of damage to the disc and even prevents the usual wear and tear. The third is that sound is recorded more than once on the disc. These features combine to improve the sound quality and the lifetime of the disc.

▼ A compact disc is 120 mm in diameter and 1.2 mm thick. It can store up to 74 minutes of music. A close-up of the surface of the disc (on the right) shows the pattern of pits which store the sounds.

In digital recording, the sound is not recorded as the depth of a groove nor the magnetization of a tape. Its intensity is converted into a binary number and this number is put on to the disc. All the zeros in the number are represented by small pits in the disc.

Reading by laser

As a laser scans the disc, its light is reflected efficiently by the smooth surface of the disc but is scattered in all directions when it strikes a pit in the surface. By measuring changes in the brightness of the reflected light, the disc player can read the lines of zeros or ones that represent the sound intensity values. By converting these intensities to electrical signals which drive a loudspeaker, the original sound is read from the disc and reproduced so that it can be heard.

The laser beam is scanned in a spiral beneath the disc by a pair of mirrors. The beam is focused on the recorded surface through the plastic cover of the disc. Lasers are used because they are very bright and can be focused on to very small spots.

To reduce the likelihood of making a mistake in reading the disc, the sound values are recorded several times on the disc. All the recordings are read by the laser and compared with each other. If one reading is different from the others because of a scratch or mistake in the manufacture, that particular reading is not sent to the loudspeaker.

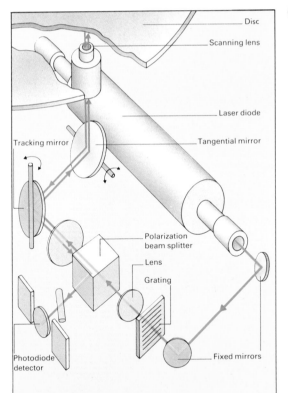

Reading from a compact disc

An array of mirrors transmits the laser beam to a scanning lens which focuses the laser light on to the underside of the disc. As the disc rotates, the lens moves to follow the spiral track of sound information. The mirrors also move to keep the beam pointed at the lens.

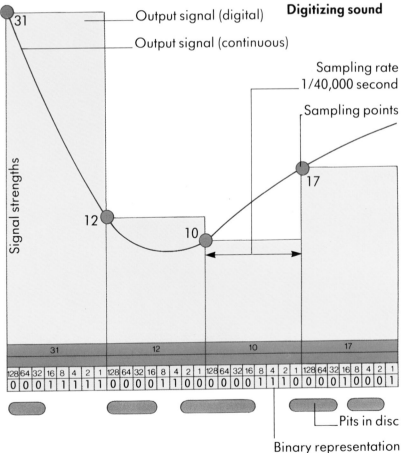

▲ Digitization of a sound signal involves measuring the intensity of the sound at regular intervals. Measurements are taken 40,000 times every second. Each value of sound intensity is converted into a binary number, a string of zeros and ones. These numbers are recorded on the disc by cutting small pits in the disc to represent a zero or a line of zeros.

Computer basics

- A "megaflop" is one million floating operations per second. Modern supercomputers operate at a speed of up to 400 megaflops.

- The ENIAC computer of the 1940s used so much electricity that the lights of Philadelphia, USA, dimmed when it was switched on.

- Work is in progress in Japan to produce a "fifth generation" of supercomputers. They would be able to perform 1,000 million logical steps per second: 10,000 times the number possible at present.

▶ Using a home computer. Computers are increasingly used in the home as well as running the operations of many businesses and industries. In the home they are used for entertainment and education; they also store information such as addresses, telephone numbers and the family accounts. Computers as word-processors are also increasingly popular.

Computers are machines that can perform three important functions. They can both store and recall information which is fed into them. The information is put into the memory of the computer and can be brought back from the memory when it is needed. They can also process that information according to certain instructions. The instructions can be fed in from the outside in the form of a computer program. Both the information and the program must be given to the computer in a form that it recognizes. Computers are becoming more powerful as it becomes possible to build smaller and smaller electronic components.

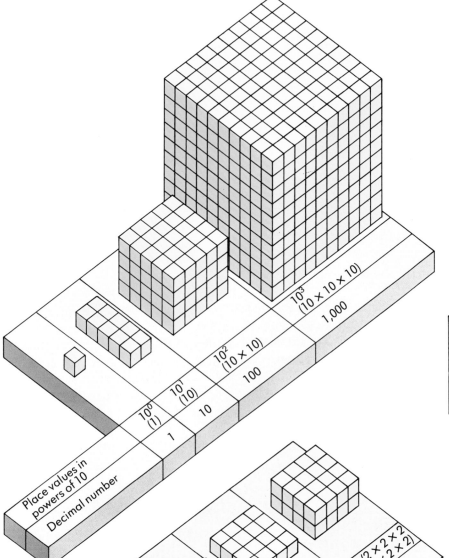

The decimal system

Our ordinary number system is based on ten numbers, or digits: 0, 1, 2, 3, ..., up to 9. It is therefore known as a decimal, or base-10 system. Numbers larger than 9 are expressed in digits using the idea of place values: the value of a digit depends on its place in the number. Place values go up in powers of ten, as illustrated on the left. (Any number to the power nought is one.) The number 153 therefore breaks down to 1×10^2 plus 5×10^1 plus 3×10^0.

Decimal number 153

Sample number	1	5	3
Power of 10	10^2	10^1	10^0
Equivalent decimal number	100	10	1

The binary system

The earliest computers carried out operations using decimal numbers. But today's computers use a number system based on two, the so-called binary system. In this system only two digits are required to express a number, the binary digits ("bits") 1 and 0. In a computer these two digits can be simply represented as, for example, the flow (1) or non-flow (0) of electric current. As in the decimal system, the binary system builds up numbers using place values. In binary the place values go up in powers of 2, as illustrated on the left. The table below shows the binary equivalent of 153.

Binary equivalent of the decimal number 153

Binary number	0	1	0	0	1	1	0	0	1
Power of 2	2^8	2^7	2^6	2^5	2^4	2^3	2^2	2^1	2^0
Equivalent decimal number	256	128	64	32	16	8	4	2	1

Microcircuits

Computers contain a very large number of switches. They are used to allow access to each element of the memory store, and also to perform the arithmetic and logic functions of the CPU. The size and weight of a computer depends on how small the switches can be made.

The modern computer can be made so small because it uses switches of microscopic size, known as transistors. These are linked in miniature electronic circuits along with other microscopic components, such as capacitors. Using such microcircuits, the entire CPU and memory of a powerful computer can take up only a few square centimetres.

Transistors and the other microcomponents are made from silicon. Silicon is a semiconductor, a material with electrical properties between a conductor and an insulator. Crystals of silicon can be made to conduct electricity a little when impurities are added to it.

A transistor is made of three layers of silicon, which have had different types of impurities added to them. The central layer, called the base, is treated (or "doped") with a different impurity from the upper and lower layers.

These are called the emitter and the collector. Small changes in voltage between the base and the emitter can switch on and off relatively large currents between the emitter and the collector. This control is very sensitive, and switching uses little power.

The microscopic components of computers are not made separately and then connected together externally. They are formed, along with the circuits that link them, within a single wafer-thin crystal, or chip, of silicon. This arrangement is known as an integrated circuit.

Hundreds of identical chips are manufactured at the same time from a round slice of silicon. The slice goes through a lengthy series of masking, etching and doping operations to create the three-layer electronic components and circuits on the chips. Masks are designed hundreds of times life-size and then photographically reduced. After manufacture, which must take place in the cleanest possible, dust-free environment, each chip is tested by probes, and the faulty ones rejected. The good ones are mounted and wired on plastic blocks for connection to external circuits.

Shrinking circuits

In the 1830s the English inventor Charles Babbage designed, but never built, a mechanical computer, called the Analytical Engine. The modern computer was born when the necessary switching became electronic, in the form of valves (vacuum tubes). The first electronic computer was built at the University of Pennsylvania in 1945. Called ENIAC (Electronic Numeral Integrator and Calculator), it used over 18,000 valves and occupied a large room. Computers became very much smaller in the early 1960s because they used tiny transistors instead of valves. Within a decade, computers had shrunk even more because of the introduction of integrated circuits on silicon chips. The shrinkage still continues: more than 1 million components can now be squeezed into each square centimetre of chip.

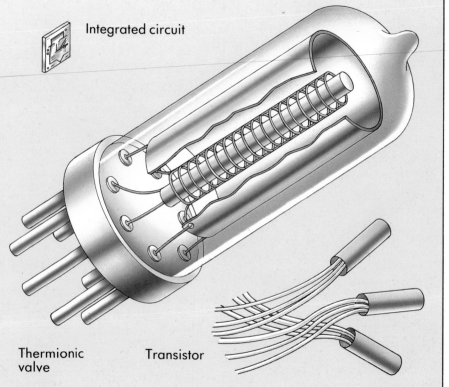

Integrated circuit

Thermionic valve

Transistor

► A very close look at the surface of an integrated circuit. In a computer, information is stored in binary code, as a series of the binary digits (bits) 0 and 1. Each bit is stored in miniature capacitors. If the capacitor is charged, that represents 1; if it is uncharged, that represents 0. The charge in the capacitor can be changed by the switching action of transistors. The transistors are turned on and off by voltage lines which run across them.

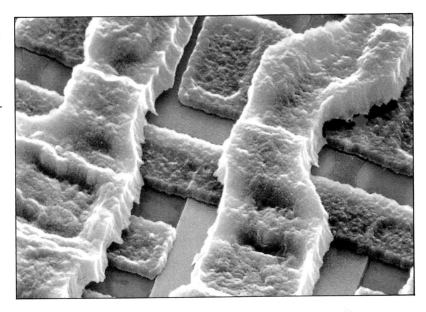

▼ A magnified picture of an integrated circuit in the jaws of an ant. Computer designers are constantly trying to produce smaller and smaller components and connections. This allows more computing power for the same area of circuit board.

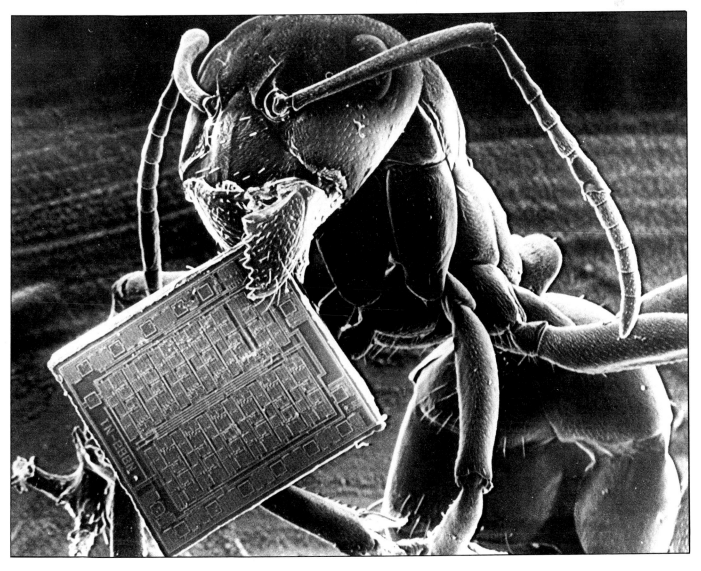

Micros and mainframes

Spot facts

• The world's first megabit (one million bit) microchip was produced in 1984. The memory was stored on a chip as thin as a human hair and the size of a drawing-pin head.

• If the fuel efficiency of the motor car had increased at the same rate as computer technology, it would run for 5 million kilometres on 4.5 litres of petrol.

• It took the personal computer just ten years to catch up with the mainframe in terms of sales. From 1977 to 1987, sales rose to $28 billion from nothing.

▶ A view inside a mainframe computer. This is the Cyber 205 supercomputer, one of the most powerful available today. It can carry out millions of arithmetical calculations every second. The wiring is extremely complex and the operation of the computer generates a high temperature. Future computers may use laser beams instead of wires to carry information.

Microprocessors and computers have caused a revolution in business, travel and communications. Microcomputers are available in many schools. Industry and commerce rely on them more and more. The power of the computers we use is increasing rapidly. A personal computer today can be as powerful as the largest computer in the world was 20 years ago. The rate of advance continues at an astonishing pace. Mainframe computers can now perform 250 million calculations in one second. No doubt even this extraordinary speed will be increased greatly in the future, as computers become more powerful.

Pocket calculators

The first mechanical calculating machines came into use in the 1600s. They were built by the mathematicians Blaise Pascal in France and Gottfried Leibnitz in Germany. They carried out calculations by means of rotating cogwheels. But mechanical calculators became obsolete with the introduction of electronic devices based on the silicon chip. The first pocket calculators appeared in 1971.

Pocket calculators have a keyboard, a processor, and a display. The processor is an integrated circuit. It contains circuits which interpret the signals from the keyboard, circuits which power the display, and two types of memory. One memory is used as a temporary store for numbers or instructions; the other is a permanent store of programs which can calculate the answer.

This is how the calculator works if the operator wants to multiply 9 × 13. When the number 9 key is pressed, the processor stores the binary code for 9, which is 1001, in its display memory. It also shows the number 9. When the × key is pressed, the processor remembers that, and also copies the display memory to another memory location, called the operand register. When 1 is pressed, the 1001 is replaced in the display memory by 0001 and the display shows 1. When the 3 is pressed, the 0001 is moved into the second display memory cell, and the first cell contains 0011. The display shows 13. If the = key is now pressed, the processor will multiply the number in the operand register by the number in the display memory and will display the answer.

Most calculators have a liquid crystal display (LCD). Behind the front screen of the display each position where a number can be shown is divided into eight segments. Seven straight-line segments are used to show all the numbers and one circular segment can be used to show the decimal point. Each segment contains a substance which can affect the properties of light passing through it. Behind all the segments is a mirror. When the display is off, light from outside passes through the liquid crystal, reflects from the mirror and back through the liquid crystal. The display is uniformly illuminated. When a small voltage is applied to the liquid crystal in a segment, it stops the light being reflected and a segment appears dark. The shapes of numbers are built up in this way. LCDs require so little power that many calculators are now powered by solar cells.

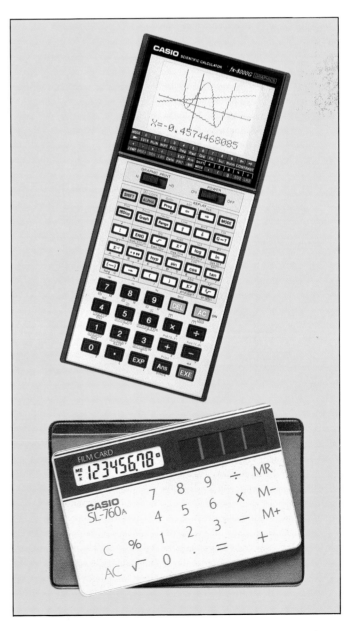

▶ Many different types of pocket calculators are now available. They vary in size and complexity. Small and simple is the so-called "credit-card" calculator (bottom), which is similar in size to a credit card. This model has simple arithmetic functions. It is powered by a set of solar cells. By contrast the scientific calculator (above) has multiple functions, including logarithms and trigonometric values. It also has a large liquid-crystal display, on which quantities can be expressed graphically. It is powered by batteries.

The micro

Microcomputers are used in all areas of business and industry, in banks, shops and offices, in the home and in schools. Knowledge and experience of computers is becoming more and more important in almost all types of work. There were particularly rapid advances in microcomputers in the 1980s because of progress in the manufacture of integrated circuits.

A microcomputer consists of a small number of chips, arranged on a printed-circuit board. Each chip is made for a particular job within the computer. The most important chip is the microprocessor. This chip contains all the circuits for performing the calculations of the computer and for checking the answers to the calculations. Other chips on the circuit board provide the memories of the computer: the ROM and the RAM. The capacity of a computer is usually expressed in terms of its RAM. A typical home computer might have a RAM of 128K. K stands for kilobytes, or thousands of bytes; 1 byte equals eight bits.

A clock chip sends out electrical pulses to various parts of the computer to ensure that events happening at different places on the circuit board are synchronized.

Peripherals
The microcomputer also contains circuits which control the flow of information into and out of the computer. These are called input/output (i/o) ports. All the devices from which the computer receives instructions, or to which it gives instructions, have to be permanently attached to the computer or plugged into an i/o port. Such devices are called peripherals.

The keyboard is an important source of input information for the computer. In most home computers the microprocessor is inside the keyboard, which also houses i/o ports.

Storage capacities compared

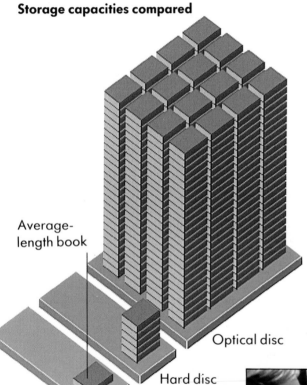

Average-length book

Optical disc

Hard disc

Floppy disc

The capacities of the different kinds of discs used to store computer data vary enormously. An optical disc can store over 200 times more information than a floppy, magnetic disc. It stores information in the same way as the compact disc used for sound recording, as a series of pits cut in the disc surface by a laser beam. The pit pattern, which represents the bits of data, can be "read" by another light beam.

▼ An engineer working with a microcomputer on a design project. He is using a "mouse" to make changes to the design he has created in colour on the screen of the VDU. Computer-assisted design (CAD) has become one of the most valuable engineering research tools.

Programs and data are usually stored on magnetic tape or disc. The information is recorded in the magnetism of crystals in the tape or disc. Floppy discs are made of flexible plastic. There are two standard sizes: 5¼ inches (133 mm) and 3½ inches (88 mm). The recordings run in circles, or tracks, on the surface. To read the disc, it is put into a disc drive, where it is rotated beneath a sensor called a head. A hard disc is built into some computers and provides much greater storage capacity than a floppy disc.

The most important output device for a computer is the monitor. This is a screen like a television screen which displays the results of the computer's calculations. The display could be words, numbers or pictures. The monitor also produces sounds for the computer. The monitor can display the output from the keyboard or disc drive which allows the programs to be written and finalized on screen before the computer is asked to print them. Some monitors allow a computer to be controlled from a touch-sensitive screen.

Peripherals

▼ Inputs to the computer can come from many different devices. The joystick lever can be moved to give direction and speed instructions, usually for a computer game. The light-pen can be used with a light-sensitive monitor to point to a particular position on the screen to give instructions. The mouse can be used to move a marker on the screen called the cursor. The cursor moves in the same direction as a ball beneath the mouse.

Plotter

Joystick

Light-pen

Keyboard

Computer

Video display unit

Mouse

Printer

Disc drive

Floppy discs

▲ In addition to the output to the monitor screen, the computer can instruct printing devices to produce a permanent record of its output. This is called a hard copy. Plotters can reproduce pictures from the monitor screen or draw graphs. Printers produce a hard copy of words and numbers. The dot matrix printer builds up the letters from dots; the wheel printer has a key for every letter, as in a typewriter. The laser printer gives the highest-quality copy.

Microsystems

The microcomputer can form the nerve centre of an extensive interacting system. The various inputs and outputs can measure and control operations, make inquiries and respond. Some methods of entering information into a computer, such as the keyboard or mouse, need a human operator. However, the microcomputer can also be set up to monitor a process or activity continuously. It can be programmed to take certain actions in response to its measurements. Computers are widely used in the chemical industry in this way. They control the quantities of the chemical substances being used, the temperature and pressure at which reactions are taking place, and the properties of the end product.

All the measurements from instruments in the factory are converted to low-voltage signals that the computer can read. The computer measures the voltage on each of its input lines in turn. If any signal is low or high compared with the numbers in the program, the computer sends control signals to the machines in the factory to put right the differences. The computer control system can run day and night with no need for supervision or rest.

► The microcomputer can communicate with many devices attached to the computer. However, using a device called a modem (modulator-demodulator) it is possible for a computer to communicate with distant machines by using a telephone line. The modem allows the computer to send and receive messages from other computers. This is called electronic mail. It is also possible for a computer to search central libraries of information called databases. There are many methods of storing instructions and data for use by a computer. The hard disc, the floppy disc and the cassette all record the information on magnetic material. With an optical disc, words and numbers are stored as a series of ones and zeros by forming a small pit to represent a zero. The disc is read by laser. Optical discs are Read-Only.

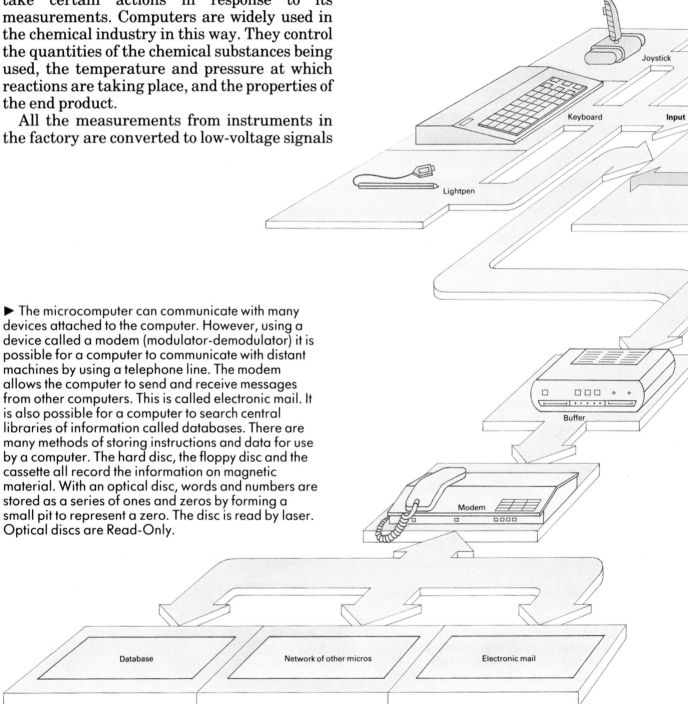

Joystick

Keyboard

Input

Lightpen

Buffer

Modem

Database

Network of other micros

Electronic mail

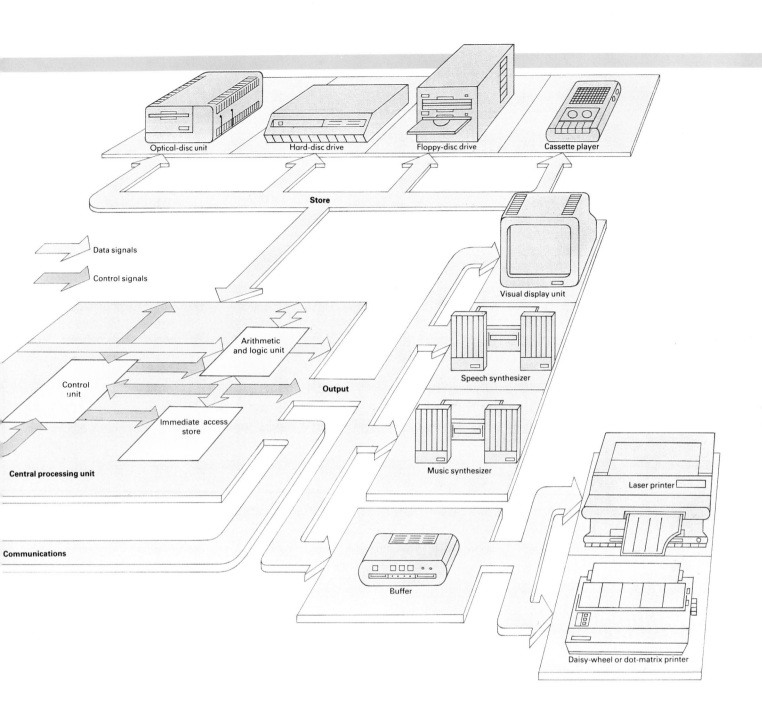

Data signals

Control signals

Optical-disc unit

Hard-disc drive

Floppy-disc drive

Cassette player

Store

Visual display unit

Arithmetic and logic unit

Speech synthesizer

Control unit

Output

Immediate access store

Music synthesizer

Central processing unit

Laser printer

Communications

Buffer

Daisy-wheel or dot-matrix printer

In order to read voltage signals from monitoring devices, a computer must be fitted with an analogue to digital (A/D) converter. The voltage measurement is an analogue signal; this means it can take any value and can change smoothly. The computer can only understand a digital signal which can be represented in binary code, and must therefore be a whole number. An 8-bit A/D converter produces the 256 binary numbers between 0 (00000000) and 255 (11111111) for an input voltage of 0-5 volts. So a measurement of 3.70 volts produces the binary number for 3.70 multiplied by 256/5, which equals 189.44. Rounded down, that is 189 (10111101).

Buffers
Computers can send out information to other devices at an extremely high rate: up to 10,000 bits per second. The receiving devices are unable to cope with the speed of the computer. It is necessary to install a unit called a buffer between the computer and the receiving device. The buffer stores the information from the computer and then sends it out at the rate that the device requires. Because the computer works so quickly, it can interact with many devices, apparently at the same time, by constantly checking all the devices to see if any of them is waiting for instructions.

Mainframes

The largest and most powerful computers are called mainframes. They are used as the central data storage and processing machines for large companies and government departments. Many other smaller computer terminals can be connected to the mainframe and all use its huge processing power.

The mainframe computers are also used to solve problems which require a vast number of calculations. One of the most complex of these is the prediction of weather based on the information gathered at weather stations. Many different physical properties form part of the calculation, such as temperature, humidity, wind speed and air pressure. All these aspects of the weather affect the others according to the laws of physics. These are fed into the computer as mathematical equations. The computer calculates how each of the properties will change in the short term and then uses these new values to predict those for the long term.

The mainframe or supercomputer can do hundreds of millions of arithmetical operations (add, subtract, multiply, divide) every second. The processing power of a computer is measured in "flops", which stands for "floating-point operations". One hundred million would be called one hundred megaflops.

Computers of this speed consume a great deal of electrical power. A major consideration in their design is how to regulate the rise in temperature generated in their operation. Some supercomputers are cooled by liquid.

There is much research being carried out into making computers even faster. "Parallel processing" allows the computer to perform many calculations at the same time in different parts of the machine. The optical computer may one day increase speeds by using light instead of electricity within the machine.

▲ A modern mainframe computer room. Because of the miniaturization of electrical components, computers have become much smaller. As a result, housing them has become easier. In the early days of mainframes, it was important that computer rooms had good air conditioning and sensitive temperature controls. Such restrictions are fewer with today's mainframes. Computer storage in this computer room is on magnetic tape, stored in plastic reels. This method is bulky, and storage may in future move to a more compact medium.

▶ A Cray machine at the US National Supercomputing Center, at the University of Illinois, USA. The computer uses parallel processing to calculate at incredible speeds, counted in billions of operations per second.

▼ A three-dimensional image generated by a powerful mainframe computer in a nuclear reseach laboratory. It shows an "event" taking place inside an electronic particle detector. The event is a collision between a proton and an antiproton, producing an electron (blue) and other particles, such as neutrinos.

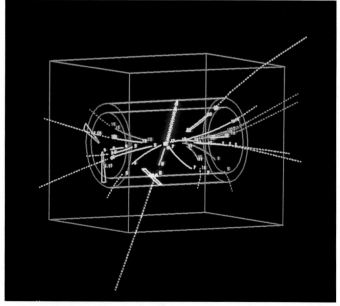

Computers at work

Spot facts

• If one in seven Americans who travel to work stayed at home and used a computer, the United States would no longer need to import oil.

• Computers are helping us do away with using cash in everyday business. When we no longer use cash, banks will save £15 million a year in moving cash around.

• A computer can be 85% as efficient as a doctor in diagnosing the cause of death in a patient. However, it is only 15% as good as a group of doctors consulting each other.

During the past ten years computers in business and industry have grown from a rarity to an essential feature of commercial life. They bring great advantages, but also some dangers. Information can be stored compactly on disc and retrieved easily, but a damaged disc can lose volumes of records. A computer memory is more easily lost than a book. Computers can communicate with each other, but this allows outsiders to gain access to the information. Updating of records is made easy, but tampering and erasing are also easy. Security of computer-based records is an important and difficult problem in today's businesses.

▶ Modern video game machines use advanced microprocessors to draw and move the graphics on the screen. Using computers, even in an arcade, is a useful experience for future employment. However, too much time spent with only a computer for company can make people addicted and antisocial.

Simulators

A simulator is a computer-controlled machine which artificially produces the sights, sounds or movement of a real-life situation. They are used to train people to respond properly in the real event.

Computer simulators and fast-action video games both rely on the ability of the computer to generate pictures and to update them quickly enough to produce the illusion of continuous motion. The speed and complexity of computer graphics has increased rapidly because of the desire of the military, aviation and space agencies to improve the effectiveness of training simulators.

Flight simulators for aircraft are essential for pilot training because of the huge risk and cost of having untrained pilots flying real aircraft. Two computers control the simulator. One operates hydraulic pistons, which lift or roll the cockpit in response to the controls. The other generates the view which the pilot sees. The computer memory contains a map of the training area, which it displays and moves according to the movement of the simulator controls.

The most recent simulators for military aircraft contain highly-detailed pictures of potential target areas. The computer can also simulate different weather conditions and create lighting appropriate for day or night. The displays for simulators have to fill the trainee's field of vision to be convincing. Lasers are used to draw the computer-generated pictures on a large screen. Another display method is to produce the image on a small screen which is incorporated into the trainee's helmet.

▼ A trainee pilot of the Swiss Air Force practising an approach to an airfield using a flight simulator. The view which the pilot sees is computer-generated and changes in response to the controls which the pilot uses. The cockpit also is moved by a computer to respond as if it were a real aircraft.

Computers in education

The use of computers in education has increased dramatically over the past ten years. Computers are now available in most classrooms for all ages of children and adults involved in learning and teaching. For young children, the computer is a source of entertainment as well as learning. Programs are available which can help children read and write. The computer will show a picture of an object or animal and ask a question about it. The child has to be able to read the question, think about the answer, and tell the computer. This is done by moving a marker to the correct answer from alternatives shown on the screen, by typing the answer, correctly spelled, or in the case of the most advanced teaching machines, by speaking to the computer.

Computers as word-processors are used extensively in education, for schoolchildren and university students. The advantages of the word-processor are so great that writers of all kinds are now using them. Students, teachers, journalists and novelists all rely on the computer in place of the typewriter or the pen. A spelling-check program can educate or merely correct mistakes, depending on the willingness of the user to learn.

▼ Computers are now used by many children at school and at home. Because the increase in the use of computers is so recent, children are often more familiar with them than their parents. The training that children receive on computers, even from playing games, will help them in their future working lives in an increasingly computerized world.

Software has been developed which allows pupils with a microcomputer to learn at their own pace, and follow their own course. The program presents the pupil with some information and a choice of which aspect of the subject they want to know more about. The pupil chooses, and that information is displayed, with another choice for further related study. Entire encyclopaedias are available on disc, containing pictures, sounds, film sequences, and copies of contemporary documents. Any of these can be chosen by selecting from a choice of symbols (called icons) displayed on the screen. Moving pictures are stored compactly by recording only the differences between one frame and the next. On videotape the whole picture is recorded for every frame.

Computers are used to teach foreign languages. The computer can show a piece of writing on screen in any language. It can ask for a translation of a word and correct any mistakes. It can ask the pupil to choose which of a choice of words best fits into the text at a particular place. Computer programs have not yet been developed which can teach a spoken language.

An important use of computers is to teach the use of computers. Programs exist which guide pupils through the process of writing computer programs. The program can monitor what is being written and point out mistakes or parts of the program which could be improved. But the main purpose of computers in schools is to give pupils practice in using them.

▼ Computers can help us communicate. The girl (inset) is using a computer which recognizes her speech when she responds to its questions. The boy (main picture) cannot move his hands, but he can use the computer by speaking into the microphone.

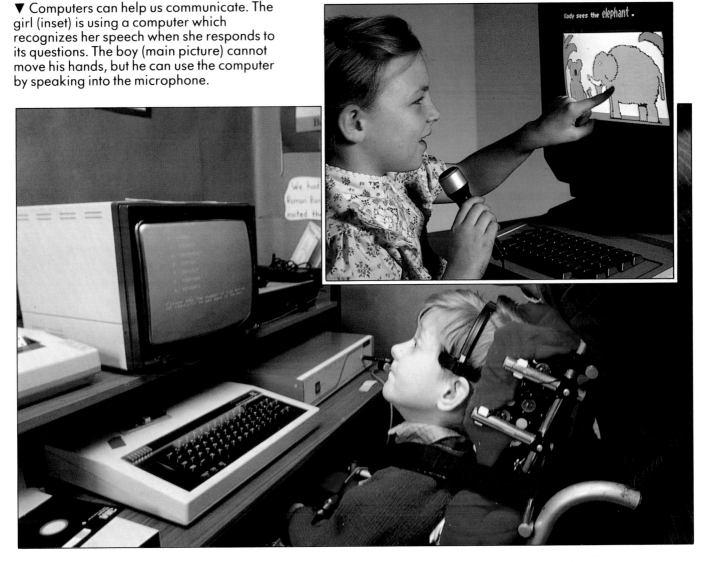

215

Computers in finance

Banks have been using computers since the invention of the electronic calculating machine in the 1950s. They have continued to use computers and automated electronic systems as the technology has become more advanced.

The use of computers most obvious to customers is in the automatic banking machines installed at most banks. Customers can withdraw cash or find out how much is in their account from the bank's records.

Many banks have joined a network which allows a customer of any bank in the scheme to withdraw money at any network bank. This system requires that the computers of each bank are able to exchange information. They must be compatible. When a customer takes out money, the computers ensure that the correct amount is deducted from the right account at the right bank.

Most international banking is now done by computer. Electronic signals travelling over telephone lines can transfer funds from one bank to another and from one continent to another. Billions of pounds are moved around the world every day by computer. Great care must be taken to keep the system secure. The computer programs which control financial business are greatly concerned with security. They must ensure that any instruction to release money is genuine.

Computers have had enormous impact on the stock exhanges of the world. At a stock exchange, dealers buy and sell shares in commercial companies. This used to be done by people with shares to sell talking to those who wished to buy. Now shares can be transferred by computer. Dealers use personal computers in the office to show them the current prices of all the shares. Prices are coloured red or blue depending on whether the price is going up or coming down. When shares are bought or sold, computers transfer the ownership of the shares and the money of the buyer. Computers can be programmed to buy or sell shares automatically when the price reaches a certain value. On bad days, many people blame the automatic nature of the computerized system for the rapid fall in share prices.

▶ The Hong Kong stock exchange. Each worker has a computer terminal, a printer, and a telephone. Nearly all communication with the outside world is electronic.

▲ A cash dispenser allows access to money 24 hours a day. Users have a plastic card and a personal number to identify themselves to the machine.

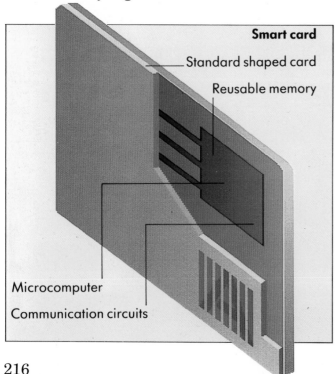

Smart card

— Standard shaped card

Reusable memory

Microcomputer

Communication circuits

◀ A "smart" card is the size of a credit card but instead of a strip of magnetic tape it has a microcomputer inside it. When used to pay for purchases, the value of the goods is automatically deducted from the value of the card.

Computers in the office

Office work is the area of business which has been most influenced by the computer and the microprocessor. Secretarial work in particular is dominated by the use of the word-processor in place of the typewriter. Printers, photocopiers and fax machines all contain microprocessors to control their operation.

The word-processor consists of a keyboard and a screen, or visual display unit (VDU). The letters and numbers of the keyboard are laid out in the same way as on a typewriter. The text is typed in and displayed on the screen. The words can easily be changed and moved about.

The word-processor can display any part of a long document, move sections of text around, change layout, produce bold, underlined or italic script, all without any retyping. When the text is complete, it can be sent to a printer, which puts the text on to paper.

Local Area Network

Letters or documents need no longer be filed in paper form. They can be filed electronically in the word-processor's memory. Standard documents can also be stored there. When they are needed, they are brought on to the screen. Names, dates or other information specific to the new document are inserted where necessary. It can then be printed.

In companies large and small computers are used to carry personnel records, calculate and print out salary information, and so on. In manufacturing industries, computers often control the purchase of equipment and materials, ensuring that the materials arrive at the right time. In sales and marketing departments computers hold lists and order requirements of customers.

Scanning system _____

Personal computer _____

▶ The modern office contains an integrated system for its information technology. Personal computers on every desk can exchange information with a central minicomputer or with peripheral devices such as scanners, copiers and printers. The system of inter-connections between the devices is called a Local Area Network.

Information processor _____

Information system _____

Word-processor _____

◀ Specialist computer systems are invaluable in design offices, as here. Using the keyboard or a "mouse" to input data, designers can easily compose and alter plans and pictures on the VDU screen. Designs can then be sent to customers via a modem and telecommunications links, or it can be faxed: sent by facsimile transmission.

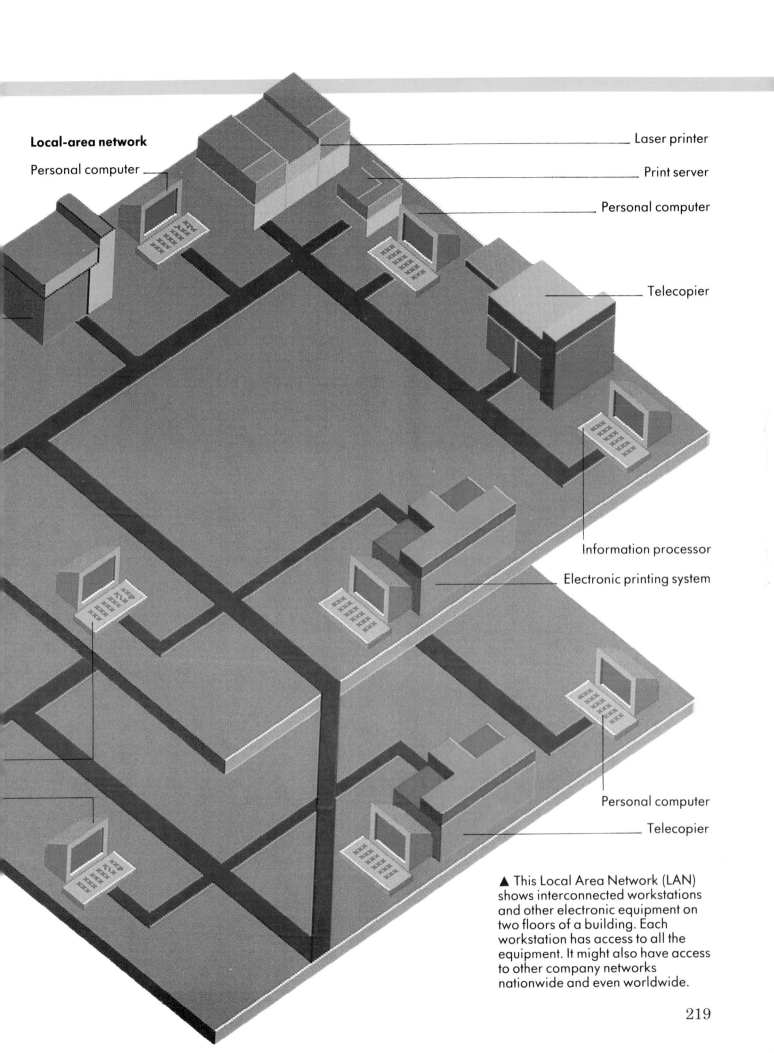

Local-area network

Personal computer

Laser printer

Print server

Personal computer

Telecopier

Information processor

Electronic printing system

Personal computer

Telecopier

▲ This Local Area Network (LAN) shows interconnected workstations and other electronic equipment on two floors of a building. Each workstation has access to all the equipment. It might also have access to other company networks nationwide and even worldwide.

219

Designing with computers

Spot facts

• Computer-aided design (CAD) could one day reduce the number of design engineers by 80 per cent.

• The first CAD video game was a simple black-and-white tennis game called Pong. It was invented by Nolan Bushnell, founder of the Atari company, in 1972.

•High resolution computer monitors can display over a million points on the screen. This is five times the resolution of a standard colour TV.

•The computing power required to store video information is daunting. An ordinary floppy disc can store only enough data to generate a tenth of a second of video film.

Everything that people make, from buildings and cars to pictures and music, has to be designed. Designing is a creative process which starts with human imagination. Successful design results in creations which accomplish the intended result in exactly the desired way.

Some computer programs are written to assist with design. With this new tool, designers can try out far more ideas than could be examined before. They can also quickly find answers to problems that previously took far too long to solve. Computers can extend the creative powers of engineers, scientists, artists and musicians.

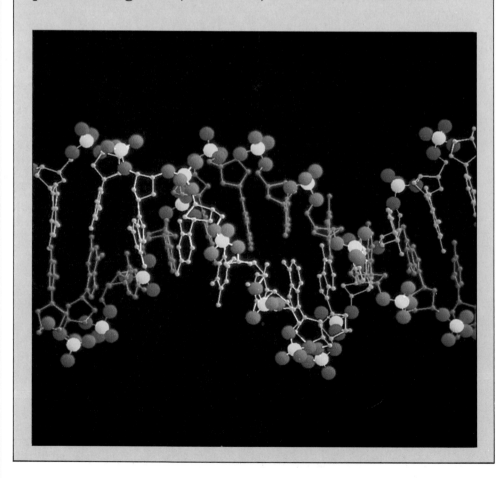

▶ A three-dimensional picture of a chemical molecule displayed on a computer screen. The image may be moved and viewed from any angle. Research scientists use the computer picture to understand the properties of a molecule. Biochemists use it to design molecules for particular tasks.

Making music

Music is constructed from many different elements. The most important ones are rhythm and harmony though there are many more. Composing music is an artistic process and, like all the arts, it has its own set of rules to guide the artist towards pleasing results. These rules are by no means rigid and most composers break them.

The idea of using machines to compose is not new. A composing machine belonging to Samuel Pepys, who lived in London in the 1600s, still survives at Magdalene College, Cambridge. Computers are good at following instructions, and programs have been made which incorporate the rules of musical composition. They also introduce some random variations to produce a different result each time. A computer was first used to compose a piece of music in 1955. Music composed by a computer is not as popular as music from a human composer. This may be because computers are unable to mimic human artistic ability. It could also be that we do not yet have programs sophisticated enough to compose well.

Synthesizers

Synthesizers are complex pieces of electronic equipment which combine a musical keyboard with a computer. A musical note has three distinct phases. In the "attack" phase the sound builds up; it then stays roughly constant, and then fades during the "decay" phase. The characteristic way in which a note goes through each of these phases gives each instrument its own distinctive sound. A synthesizer can adjust each of the phases independently, and so can produce sounds which copy different instruments. The computer produces the music played on the keyboard so that it sounds as though it comes from the chosen instrument.

▼ The contemporary band Kraftwerk, from Germany, use synthesizers extensively in their music. The musicians program them to produce exactly the sound they want. This may be varied while they play the keyboard. Interesting music is a blend of sounds, some following and some breaking the rules of harmonic theory. Bands like Kraftwerk use a computer to produce some of the music but use their talents to compose other parts.

Artistic design

The process of skilfully and imaginatively creating a new and often attractive object is called art. Computers now assist in many branches of commercial art. Graphic designers use computers extensively to plan their work. Typically, such programs allow the artist to rough out a design on the screen, trying out colours, altering lines and text at will. During the designing process, the artist may move elements about and change the picture as often as necessary. When the work is satisfactory, it can be printed out. To do this, another program converts the numbers which created the image on the screen into the string of instructions necessary to drive a colour plotting machine.

Textile designers use a similar method to allow their customers to view patterns without the need to produce fabric. Once the plotted pattern is agreed, the computer generates the sequence of instructions necessary to drive a weaving machine, and the pattern is automatically produced.

Computers are also used to renovate valuable old paintings. Details of the painting are first entered into a computer as a very long string of numbers representing the colour at each point in the picture. The program already has a knowledge of how different renovation techniques affect each colour. The computer then calculates how the painting would look if a particular renovation technique were used on it. The renovator selects the best process.

Computers also assist the cosmetic industry. Exact colour production is very difficult but becomes extremely important where make-up is concerned. Computers help beauticians define and produce exactly the shade they want.

▼ An artist (below left) painting with computer graphics, using the screen as his canvas. He selects a colour from the multi-hued "palette" on the right of the screen and directs it to the desired place in his picture. A beautician (below right) uses a computer-aided design program to picture realistically the effects of different colours of make-up.

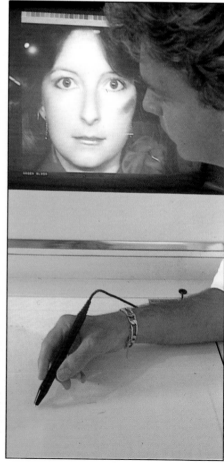

Computer graphics provide many new opportunities for artists. The "palette" from which a computer artist can select colours allows for any shade to be moved easily and quickly. Areas of the picture may be repeatedly coloured over and still appear fresh, and there is no waiting for paint to dry. The computer can produce a string of new pictures exactly like the original. Each one can then be altered slightly, so that when replayed in sequence, they give the impression of continuous movement. This technique is called computer animation.

Computers can also be used to produce abstract artistic patterns. Such programs contain rules which produce a variation in colour over the picture. There are no programs that actually make up a picture the way an artist does. As with music, this is not because computers cannot physically do what is necessary, but because people cannot write down what instructions must be followed in order to produce an original and pleasing result.

▲ A graphics designer uses a computer and "painting" software to experiment with different patterns and colours to create unusual visual effects.

▼ Computer graphics programs give commercial artists new dimensions. It enables them, for example, to create unusual typefaces for advertisments.

Computer imaging

Computers can be used to display as a picture the measurements taken in experiments. The picture is much easier to understand than the long string of numbers used to make the picture. This is called computer imaging. For example, the experimental measurements might be the amount of rainfall at different places in a country or the intensity of X-rays passing through different parts of a patient's body. The results are shown as a map.

Usually a colour code is used to show areas at which similar results were obtained. The changes in colour show the variations in the measured values. The presentation of the measurements as a multi-coloured picture can immediately give an impression of the shape and sometimes even the motion of objects invisible to the human eye.

Computer-imaging techniques are required to display pictures of the ground taken by remote-sensing satellites, such as Landsat. The satellite records electronically the intensity of reflected light at a number of wavelengths, and sends back signals that computers process into false-colour images.

A computer display can be used to picture things which cannot be photographed. For example, astronomers measure the intensity of invisible radiation, such as radio waves coming from stars, and use a computer to display the results. The results of radar detection of aircraft or of weather formations are also displayed on a screen.

Computers are also used to interpret sounds used as underwater detectors. Sound waves are sent down from a surface vessel. The strength of the echo is measured, and the results are displayed as a map of the seabed. This method can detect sunken ships and was used to find the *Titanic*. It has also been employed in the underwater search for the legendary Loch Ness monster in Scotland.

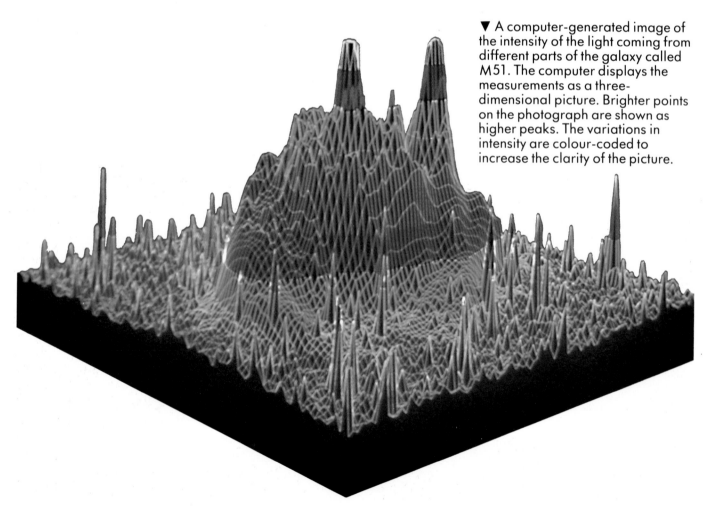

▼ A computer-generated image of the intensity of the light coming from different parts of the galaxy called M51. The computer displays the measurements as a three-dimensional picture. Brighter points on the photograph are shown as higher peaks. The variations in intensity are colour-coded to increase the clarity of the picture.

Computer imaging is extensively used in hospitals to create pictures of the inside of a patient's body. These reveal internal problems without the need for an operation. High-frequency sound called ultrasonic sound is used to picture unborn babies inside their mothers. Doctors can check that the baby is developing properly.

Radio waves are used to study the chemicals in the human body by magnetic resonance inspection (MRI). The patient lies inside a large cylindrical magnet. In a magnetic field, atoms are able to absorb radio waves of a certain frequency. Different atoms absorb different frequencies, so the concentration of each kind of atom can be measured within the body.

The most powerful imaging method is called computer-assisted tomography (CAT) scanning. A flat beam of X-rays spreads out from a source, passes through the patient, and is detected behind the patient. By rotating the X-ray source, the section of the body being examined can be viewed from every angle. A computer then composes a complete picture.

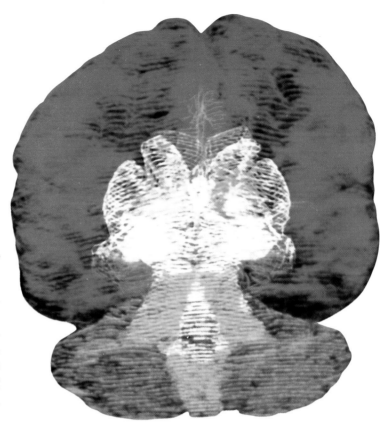

▼ A kaleidoscopic false-colour image of the Newquay area of Cornwall, England, produced by computer-processing data acquired by a Landsat satellite. It shows the patchwork pattern of fields in various hues. The prominent linear feature in the middle of the image is the runway of Newquay airport.

▲ A picture of a human brain as seen from behind. Different areas of the brain are assigned colours by the computer. The red areas represent the hemispheres of the forebrain, which control thought, memory, sight and language. The orange areas control balance and muscle movement.

Molecular modelling

Molecules are built up of atoms. There are only about one hundred different types of atom, but they can be combined in countless different ways to form a huge variety of molecules. Strict rules govern the ways in which atoms may combine with each other. This means that computers can be programmed to model them.

Researchers into molecular structure often know which atoms are present in a molecule, without knowing the molecule's shape. Many of the chemical and biological properties of a molecule depend crucially on its shape. A computer may be used to examine the possible molecular structures which it does by applying the rules of atomic physics. The computer can suggest the most probable shape for the molecule.

Doctors who are researching into the causes of disease often use biochemistry to isolate the particular molecules which are responsible. They can then produce a computer model of the molecular shape. Using the computer again, they can design new molecules which have the right shape and properties to interfere with the effects of the harmful molecule. These programs give doctors an important tool to help their search for new drugs and treatments for illness.

▼ A computer-generated image of the influenza virus. The central core contains the DNA molecules. The projections on the virus help it break into a healthy cell. Once inside the cell, the virus reproduces rapidly. The computer modelling of the virus can help in the design of chemicals to restrict the damage it can do.

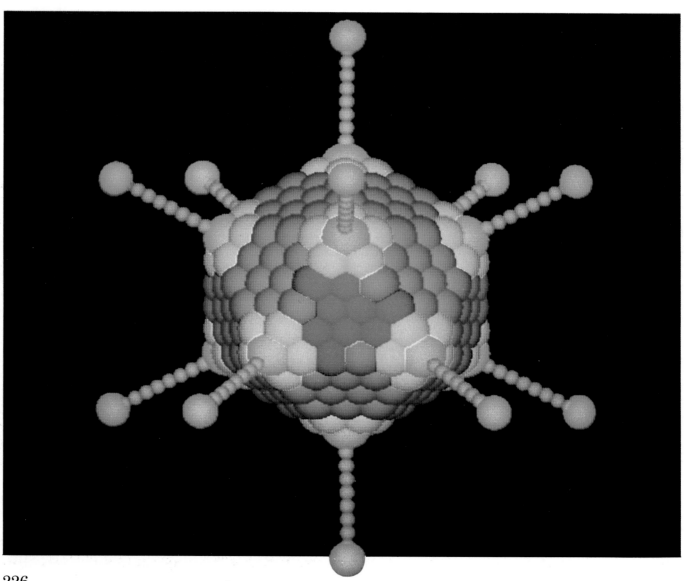

The modelling and display of large biological molecules such as proteins requires enormous computing power. The relative positions of thousands of atoms have to be given to the computer. Each different type of atom is shown in a different colour and as a different size. The computer then calculates which atoms are in the foreground and which are partly hidden by other atoms. Finally, on each atom the computer adds shading to make each circle appear to be a sphere.

▼ Scientists now use computers to study the molecules of existing compounds and to help them design new ones. The molecules are imaged in 3D and can be modified on the screen (inset). The main picture shows an image of a molecule of haemoglobin, the substance that gives blood its red colour.

Viruses that cause diseases such as influenza and AIDS can be visualized using computer modelling. The models help doctors to design vaccines against viruses. Unfortunately, viruses change, so that within a few years it may be necessary to design a new vaccine against the modified virus.

Computers can help in understanding molecular structure by automating experiments and analysing results. The molecule which is the basis of genetic material is called DNA. Many people are working on the analysis of human DNA, which consists of a long chain of thousands of millions of molecules of only four different types. Discovering the exact sequence of these four molecules would take thousands of years to achieve without computers.

Computer-aided design

Engineers imaginatively create appealing new devices which fulfil their desired function. This process involves calculations, examination of alternatives, and the analysis and presentation of solutions. Computer-aided design (CAD) assists engineers in all of these tasks.

CAD programs usually allow the engineer to enter a representation of the design through a movable hand-held device called a mouse. The program converts the designer's instructions into numbers, which are processed to represent the object on a screen. There are 3D CAD programs which allow this picture to appear to be in three dimensions (3D). The engineer may view the device from any angle and so have a very clear idea of how the object will look and work. When satisfied with the concept, the engineer must produce detailed drawings and a component list so that the device can be made. Two-dimensional programs are used for this, and take their inputs either from 3D programs or directly from the designer.

Computer-aided design programs on desktop computers have replaced the drawing board completely in many industries. When these drawings are completed on screen, another program is used to convert the numerical representation of the drawing into instructions for a printing machine to follow. Other prog-rams are used to analyse the components, and predict how each will behave. For example, the movement in a structure carrying a load can be calculated. A program can plot the way in which heat or electricity will flow through the component, and show how much it will cost to make.

It is possible to integrate several CAD systems with manufacturing machines to form a complete process from design to production. Examples of this are rare because of the expense and complexity of introducing it, but in some industries it may be necessary for commercial success. Car manufacturers in Japan, for example, already use such systems. Designers produce sketches of an idea from which detailed drawings are produced. The computer uses its knowledge of existing parts, supplies and machines to produce a programme of work necessary to make the design. When a machine is ready to produce each component, the system gives the necessary sequence of instructions directly to the machine. When all the parts have been prepared, the computer informs the engineers who may then arrange for assembly. The use of computer-aided design programs cuts down the cost of product development. It reduces the number of prototypes which have to be made and then scrapped because they are not quite right.

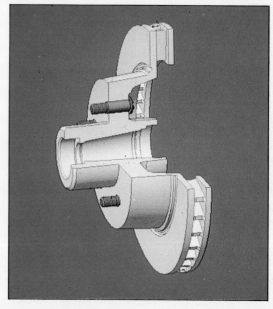

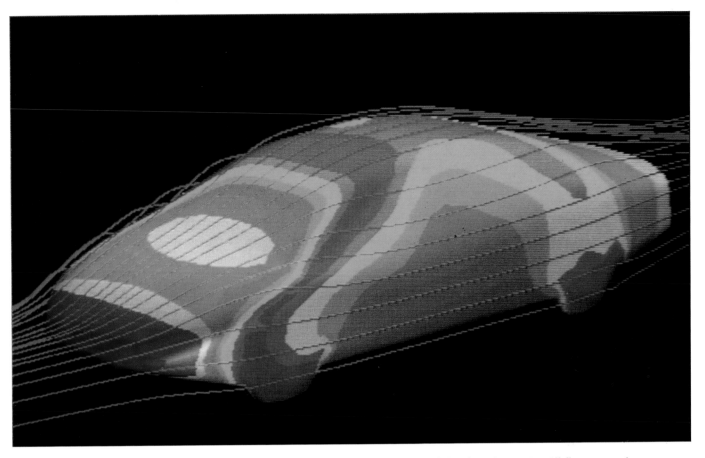

▲ New cars are analysed with CAD to produce an optimal body shape. Here a complex computer imaging package takes the results of a large design program and displays how air will flow over the car. Colours show how fast the air is flowing at the car's surface and hence how aerodynamic is its shape.

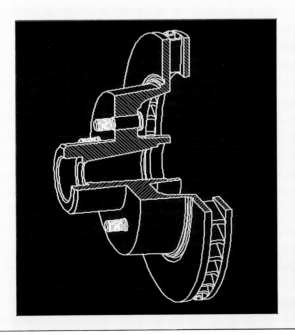

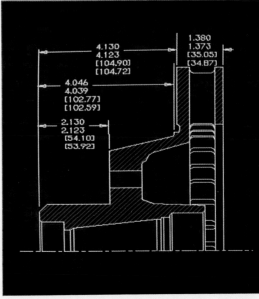

Engineering CAD

This disc brake rotor assembly is first drawn as a profile. The computer displays it from all angles. The operator can alter one measurement, and the computer changes all the others to match. It adds shading to make the drawing look more solid. All the measurements are stored in the computer's memory and can be used as a pattern during manufacture.

Robot control

Spot facts

• *Robotized trains run in the Bay Area of California, USA. At peak times, 100 driverless trains run at high speed with only a 90-second interval between them.*

• *The automatic pilot of an aircraft is a typical "invisible" robot. An aeroplane may be landed by automatic pilot with a chance of failure of one in ten million.*

• *In 1988 there were 175,000 industrial robots in Japan. There were less than half that number in the rest of the world.*

• *The earliest kind of robot was a water-clock, called a clepsydra, which was invented by Ctesibus of Alexandria, Egypt, in about 250 BC. It recycled its water by means of a siphoning device which ran automatically.*

A robot is a machine that can be instructed to perform tasks which would otherwise have to be done by people. Most robots have computer "brains", which can be programmed by an operator to carry out specific tasks. The latest robots are equipped with sensors which enable them to see, hear, touch and even smell their surroundings. They have an electronic brain sophisticated enough for them to decide how to act in response to this data. Hundreds of thousands of robots are now at work, and in future, with advances in computer technology, robots will have even more complex "brains", which will enable them to learn from experience.

▶ This Japanese robot has been taught to play an electronic organ. In its head is a television camera which enables it to see where the keys are so that it can direct its fingers to the right notes on the organ.

Automation

Before 1920 no one had ever heard of robots. The word was coined by the Czechoslovakian playwright Karel Capek, from the Czech words "robota" meaning slave labour and "robotnik" meaning slave. In his play *R.U.R.* (Rossum's Universal Robots), robots were intelligent, hard-working, human-shaped machines which rebelled against their human masters and killed them when they were used for making war. Since then robots have featured in many books and films, sometimes as good and helpful, but often as evil and destructive.

It is their ability to be instructed which distinguishes robots from machines that imitate human actions by mechanical means. Two thousand years ago, the Greeks and Romans could make statues that moved by hydraulic power. In the 1700s and 1800s there were life-size clockwork models which could even write and play musical instruments. However, these were only mechanical toys, and the correct name for them is "automata".

As long ago as 1804, the French engineer Joseph Marie Jacquard invented a loom which could be programmed to weave different patterns in silk cloth. This was done by inserting one of a series of stiff cards punched with holes into the loom. The card prevented some rods from carrying thread into the weave of the cloth while allowing others to pass through the holes. There were 400 of these weft threads altogether so that very intricate patterns could be created. A revolutionary feature was that one machine could follow many different programs. Equally, many machines could be programmed to produce identical patterns. This was the first automation of an industrial process.

In the 1950s machines came into use that could operate automatically, following instructions on punched paper tape or magnetic tape. These and later computer-controlled machines ushered in increasing automation in industry, bringing about a second industrial revolution. In the last few decades tireless computer-controlled robots have taken their place alongside human beings on factory production lines.

▼ A robot artist demonstrates its versatility at an industrial exhibition in Japan. This country boasts more robots than the rest of the world put together.

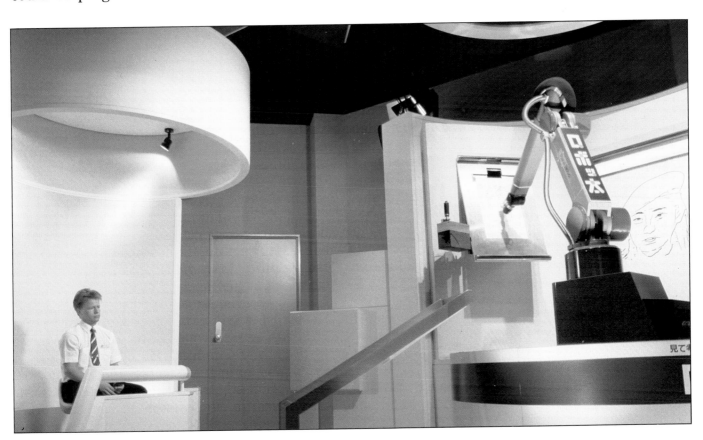

231

Designing robots

The human body is like a superbly engineered machine controlled by an incredibly complex computer: the brain. It will be a long, long time before robots are capable of responding to their environment, learning from experience and thinking creatively with the dexterity of the human brain. Arms and hands are the human limbs whose functions are most often duplicated by robots. To have the necessary freedom of movement, robot arms, like human ones, have joints capable of moving independently of one another, up and down, side to side, and in and out. Robot arms can do some things better than ours, such as picking up heavy objects with ease, or rotating their "wrists" in a full circle.

The robot must be given a computer "brain" which will assess the difficulty of the task it has been given and decide how to set about achieving it. Suppose that a robot needs to pick up an object from a conveyor belt and place it in a box (a relatively simple task for a human).

To do this it needs an electronic "eye" to identify the object; if the object is not lying in the expected position, the "eye" relays this information to the computer, which reacts by instructing the arm to approach the object from a different angle to pick it up.

Many of today's robots can use "feedback". This is the ability to make decisions which depend on information about changing conditions. We have nerves to carry instructions from our brains to our muscles. Robots have electronic cables which transmit instructions from the computer to its motor-driven parts.

It is rare for robots to walk upright on two legs. This is because walking on two legs involves lifting one foot from the ground at every step, and becoming unstable. To give a robot the balance and coordination necessary to walk on two legs would take up a wasteful amount of space in its computer brain. So most mobile robots move on wheels, or on four or six legs.

▼ To hold this egg without breaking it, the robot hand needs information about the egg's weight and the fragility of its shell. Sensors transmit this information to the robot's computer "brain", which controls the strength and precision of the hand's movements.

▼ Pictured here are the plans and parts needed to build a robot arm and hand. There are sensors to gather information about the outside world, mechanical parts for the arm itself, and the connections to the controlling computer.

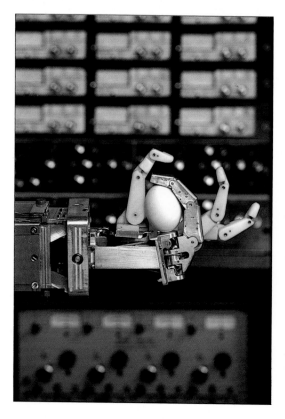

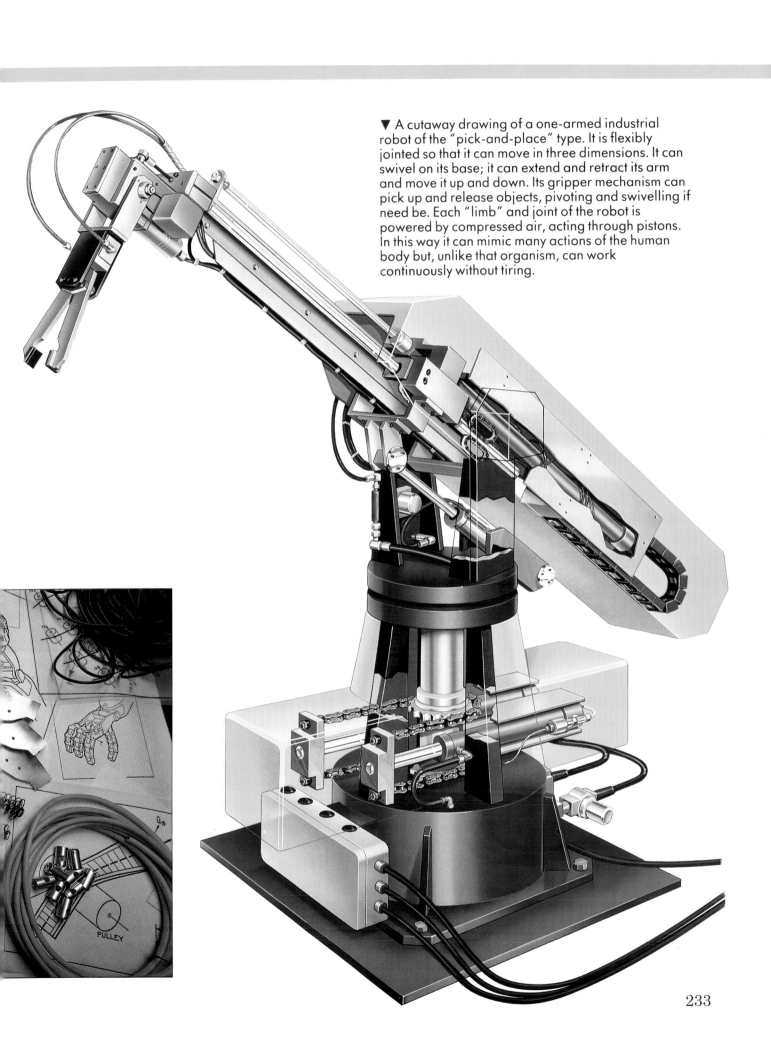

▼ A cutaway drawing of a one-armed industrial robot of the "pick-and-place" type. It is flexibly jointed so that it can move in three dimensions. It can swivel on its base; it can extend and retract its arm and move it up and down. Its gripper mechanism can pick up and release objects, pivoting and swivelling if need be. Each "limb" and joint of the robot is powered by compressed air, acting through pistons. In this way it can mimic many actions of the human body but, unlike that organism, can work continuously without tiring.

PULLEY

233

Robots at work

Robots now do many jobs which used to be done by people. They are particularly useful at industrial jobs which are repetitive, boring or unpleasant. Examples are riveting and welding in assembly plants, and spraying paint on to car bodies. Robots are better at this sort of job than people, because they can always work at the same level of precision and accuracy, and they never tire. Their work is therefore always of the same quality, and they can do more of it because they do not need rest breaks.

Robots can also do jobs which are dangerous for human beings. They can handle radioactive materials in nuclear power stations, or toxic chemicals, without needing protective clothing. They can also work in environments which would be too hot or too cold for humans. They can be used where a human life would be at risk, such as disarming bombs or searching for explosives, and of course in space.

Robots are ideal for space work because they do not need air to breathe and never grow old. Robots are used to carry out maintenance work on Earth satellites and to travel to distant planets for purposes of exploration and discovery, like the Voyager space probes.

Robots as carers

An increasing number of robots are being used as domestic help, particularly to provide services for physically handicapped people. A robot helper can enable handicapped or frail elderly people to live alone, independently of their families, and this saves them having to move into a hospital. Research is now being carried out, particularly in Japan, to develop robots capable of giving safe nursing care to sick or elderly people, and this is expected to be an area of rapid growth in the near future for a new generation of very intelligent robots.

▶ A robot waiter serves food in a restaurant in California, USA. Robots are already perfectly capable of performing simple domestic tasks like this. Robots are now being developed for more complex and demanding work, such as the care of the elderly and handicapped.

▼ An artist's impression of the *Voyager 2* space probe approaching the planet Uranus in January 1986. The probe transmitted its observations back to Earth as radio waves via the dish aerial, which had been programmed always to point in the right direction.

Index

Picture Credits

b = bottom, t = top, c = centre, l = left, r = right
ARPL Ann Ronan Picture Library. FSP Frank Spooner
Pictures. MARS Military Archive and Research
Services. MH Michael Holford. NHPA Natural History
Photographic Agency. OSF Oxford Scientific Films. PP
Picturepoint Ltd. PEP Planet Earth Pictures. RF Rex
Features. RHPL Robert Harding Picture Library. RK
Robin Kerrod. SC Spacecharts. SCL Spectrum Colour
Library. SP Sci Pix. SPL Science Photo Library. TSW
Tony Stone Worldwide.

10 Zefa/R Smith. 12 SCL. 13b Zefa/G Kait. 14 RHPL.
15t SP. 15b Colorific/J Howard. 16 Zefa. 18 SP. 20-21
RHPL. 22l Zefa/D Cattani. 22r RK. 24 SP. 25 RF. 27
Zefa. 28 Zefa. 29 RF. 29 inset Associated Press. 33 SP.
34 Zefa/W Mohn. 36 RF. 37 SPL/Tom McHugh. 38 SC.
39 FSP. 40 Zefa/E Christian. 41l, 41r Zefa. 42 CEGB.
44 SPL/Tom McHugh. 44-45 SPL/Lowell Georgia. 46
RK. 47 Zefa/J Pfaff. 48 Zefa/Streichan. 49l ARPL. 49r
Michael Holford/Royal Institution. 50 Zefa. 52-53
Quadrant Picture Library/*Nuclear Engineering*, 55l G R
Roberts, 55r Jerry Mason/*New Scientist*, 56bl SPL/Hank
Morgan. 56br SPL/US Dept of Energy. 57t PP. 57b
Zefa/J Pfaff. 58 Zefa. 59l ARPL. 59r Michael Freeman.
61 PP. 63t RK. 63b Panasonic. 66 RK. 67l The
Ridgeway Archive. 67r ARPL. 68 John Watney Picture
Library. 70 All Sport/Vandystadt. 71 Ford Motor
Company. 74-75 SCL/D J Heaton. 76 SP. 77l Quadrant
Picture Library. 77r SP. 78 Zefa/E M Bordis. 79t SC.
79cr SC/NASA. 80l RK. 80r NASA. 83tl, 83tr SC. 83b
MARS/US Army. 84l ESA. 84r SC. 85 SPL/Novosti.
86, 87 SC. 88t SC. 88b RF. 89l SC. 89r Zefa/Damm. 90,
91l 91r, 92l, 92r SC. 95 Nippon Steel Corporation. 97t,
97br Colorific/Terence Spencer. 98 BASF. 99
Zefa/Orion Press. 100-101 Bayer. 102c SC. 102b Zefa.
103 Hoechst. 105 Du Pont (UK) Ltd. 106 Zefa/Corneel
Voight. 107c Holt Studios. 107b FSP. 108 Distillers
Company. 109 Zefa. 111 International Flavours &
Fragrances Inc. 112tl SPL. 112tr Explorer. 113 Celltech
Ltd. 114 Zefa/Horrowitz. 117l Michael Holford. 117r
Zefa/Simon Warner. 118l, 118r SC. 119l Zefa. 119r RF.
120l SPL/Darwin Dale. 120-121 TSW. 121cr
OSF/Manfred Kage. 121bl TSW. 121br SC. 122
Zefa/APL. 127 SP. 131 BMW. 132-133, 132bl
Lawrence Clarke. 133t, 133b, 134 SP. 136t Union
Pacific. 136-137b SP. 138, 139, 140t, 140b RK. 141l
French Railways Ltd. 141r SP. 142 Zefa. 144, 145l RK.
145r Zefa. 146 British Hovercraft. 147t MARS/Boeing
Marine Systems. 147b SP. 148-149b, 149b PEP/John
Menzies. 149t Colorific/Dick Halstead. 150 Du Pont
(UK) Ltd. 151t, 151b RK. 152 SC. 154 Colorific/Doug
Shane. 154-155 SC. 155t RK. 155b Beech Aircraft
Corporation. 156l Telegraph Colour Library/J Reardon.
156r, 157 RK. 158, 159l SC. 159r Shell (UK) Ltd. 160
The RidgewayArchive. 161l RF. 161r, 162, 163 Zefa.
164l Mary Evans Picture Library. 164r Art Directors.
166 SCL/Muhlbergr. 167l Bayerisches National
Museum. 167r ARPL. 168 Kodak. 171t MH/Science
Museum. 171b Minolta. 172l NHPA/S Dalton. 172r
SPL. 173 PP. 174 Sally & Richard Greenhill. 175l, 175r

MH/Science Museum. 176l British Telecom. 176r
Telefocus. 178 AT&T. 179t Telefocus. 179b John Price.
180 RK. 182 SP. 183 The Marconi Company Ltd. 184
John Walmsley. 185t PP. 185b SPL/Jerome Yeats. 187
Paul Brierley. 188 Zefa/Tom Tracy. 189l RF/Images by
Goodman. 189r SP. 189b RF. 190 PP. 192t SCL. 192bl
ARPL. 193br SPL/Dr Jeremy Burgess. 194 PP. 195
Canon. 196l Sony. 196r Margaret Coreau. 198 Zefa. 199
SC/NASA. 200l, 200r Sally & Richard Greenhill. 203t
SPL/D Scharf. 203b University of Liverpool. 204
Rainbow/Hank Morgan. 205t, 205b Casio Electronics
Co Ltd. 206l Imagine. 206r Zefa/Stockmarket. 210r
SPL. 211 Woodfin Camp & Associates/M L Abramson.
212 Sally & Richard Greenhill. 213, 213 Inset
Rediffusion Simulation Ltd. 214 FSP/Mega. 215b Pete
Addis/*New Scientist*. 215r SPL/Hank Morgan. 216l
Imagine. 216r IBM/Eurocoor. 217 Art Directors. 218
IBM. 218-219 Imagine. 220 Chemical Designs. 221l
London Features International Ltd. 222 COI. 222r
SPL/Hank Morgan. 223t SPL/Los Alamos National
Lab. 225t SPL/Petit Format/B Livingston. 225b SC. 226
SPL/Division of Computer Research & Technology,
National Institute of Health. 227l, 227r Chemical
Designs. 228l Integraph (CB) Ltd. 228r, 229bl, 229br
Applicon (UK) Ltd. 229t SPL/Hank Morgan. 230
FSP/Gamma. 231 SCL. 232, 232-233 Rainbow/Dan
McCoy. 233 Sci Pix/Hall Automation. 234 NASA. 235
SPL/Lowell Georgia.